Strategies for Sustainability

Series Editors

Rodrigo Lozano, Faculty of Engineering and Sustainable Development
University of Gävle
Gävle, Gävleborgs Län, Sweden

Angela Carpenter, Faculty of Engineering and Environment
University of Gävle
Gävle, Sweden

The series focuses on "implementation strategies and responses" to sustainability problems – at the organizational, local, national, and global levels.

Our objective is to encourage policy proposals and prescriptive thinking on topics such as: sustainability management, sustainability strategies, lifestyle changes, regional approaches, organisational changes for sustainability, educational approaches, pollution prevention, clean technologies, multilateral treaty-making, sustainability guidelines and standards, sustainability assessment and reporting, the role of scientific analysis in decision-making, implementation of public-private partnerships for resource management, regulatory enforcement, and approaches to meeting inter-generational obligations regarding the management of common resources.

We favour trans-disciplinary perspectives and analyses grounded in careful, comparative studies of practice, demonstrations, or policy reforms. This largely excludes further documentation of problems, and prescriptive pieces that are not grounded in practice, or sustainability studies. Philosophically, we prefer an open-minded pragmatism – "show us what works and why" – rather than a bias toward a theory of the liberal state (i.e. "command-and-control") or a theory of markets. We invite contributions that are innovative, creative, and go beyond the 'business as usual' approaches.

We invite Authors to submit manuscripts that: - Document and analyse what has and has not worked in practice; - Develop implementation strategies and examine the effectiveness of specific sustainability strategies;

- Propose what should be tried next to promote greater sustainability in natural resource management, energy production, housing design and development, industrial reorganization, infrastructure planning, land use, business strategy, and organisational changes; - Prescribe how to do better at incorporating concerns about sustainability into organisations, private action, and public policy;

- Focus on trans-disciplinary analyses grounded in careful, comparative studies of practice or policy reform; and

- Provide an approach "...to meeting the needs of the present without compromising the ability of future generations to meet their own needs," and do this in a way that balances the goal of economic development with due consideration for environmental protection, social progress, and individual rights.

Themes covered in the series are:

Sustainability management
Sustainability strategies
Lifestyle changes
Regional approaches
Organisational changes for sustainability
Educational approaches
Pollution prevention
Clean technologies
Multilateral treaty-making
Sustainability guidelines and standards
Sustainability assessment and reporting
The role of scientific analysis in decision-making
Implementation of public-private partnerships for resource management
Governance and regulatory enforcement
Approaches to meeting inter-generational obligations regarding the management of common resources

More information about this series at https://link.springer.com/bookseries/8584

David Rodríguez-Rodríguez
Javier Martínez-Vega

Effectiveness of Protected Areas in Conserving Biodiversity

A Worldwide Review

Springer

David Rodríguez-Rodríguez
European Topic Centre - University of Malaga
Malaga, Spain

Javier Martínez-Vega
Spanish National Research Council
Institute of Economy, Geography and Demography
Madrid, Spain

ISSN 2212-5450 ISSN 2452-1582 (electronic)
Strategies for Sustainability

ISBN 978-3-030-94299-1 ISBN 978-3-030-94297-7 (eBook)
https://doi.org/10.1007/978-3-030-94297-7

This Springer imprint is published by the registered company Springer Nature Switzerland AG
The registered company address is: Gewerbestrasse 11, 6330 Cham, Switzerland

Photo: Archipelago Chinijo Nature Park, Spain. Credit: David Rodríguez

To all the anonymous heroes who lost their lives protecting biodiversity

Humble memorial to park rangers assassinated by poachers.
Niokolo Koba National Park, Senegal.
Credit: David Rodríguez

Foreword

The world faces the catastrophic destruction of nature caused by human activities and their consequences including habitat loss and the irreversible impacts of climate change. Most global frameworks striving to address this identify the dependence of people on the natural ecosystems that are the foundation of society and all economy, and argue for the protection and wise use of nature through protection; conservation, including wise use; and restoration. At the heart of these responses is the governance and management of places, to protect and conserve areas that maintain the ecosystems, species and genetic diversity of intact nature in perpetuity. The scientific literature abounds with examples of how protected and conserved areas are among the most effective mechanisms to ensure the long-term persistence of nature, and this has resulted in ambitious targets for area-based conservation being agreed and proposed among the range of other approaches, which must also include addressing the main drivers and pressures of nature's destruction. Many nature conservation agencies and indeed decisions within the international conventions, such as the Convention on Biological Diversity, focus on the management of sites that are designated as protected and conserved areas, with a view towards measuring whether their management is effective. In this volume, the authors have explored the evidence for management effectiveness diligently, to better understand which factors lead to well-managed protected areas, and whether or not these approaches have had the desired results. Implicit to this analysis is whether the purpose of each protected area is achieved, and whether or not the conservation is effective in achieving the desired outcomes. The answers are by no means straightforward, and the extensive literature review and analysis of studies presented here have focused on this question and reveal its complexity. The insights reveal that protected areas indeed do better than the surrounding landscapes/seascapes in conserving nature; however, this is conditioned by multiple variables both within and beyond the sites themselves. The reader will find this extensive review and findings a fascinating and ready resource for better understanding the circumstances and opportunities to achieve better outcomes.

Whether outcomes are achieved is largely a factor of whether governance and management are focused on the key variables related to the purpose for which the area was established. In the many techniques for undertaking protected area management effectiveness assessments, this is often weakly understood. If an area's significant values are not identified, then it is unlikely that the threats to these values can be addressed. If the significant values cannot be measured and their status and trends monitored over time, then it is unlikely that governance and management will have the critical information on which to act. But the whole enterprise of measuring and managing for effectiveness has not always achieved maturity either conceptually or in practice. Firstly, there is a blind spot.

Many of the world's natural ecosystems persist to the present day because they are actually governed and managed effectively by a wide range of actors and authorities, who did not explicitly designate the areas to be protected areas, nor explicitly state management objectives and outcomes that are amenable to techniques such as management effectiveness assessment. The territories and areas conserved by indigenous peoples and local communities in many parts of the world are the embodiment of cultural and spiritual associations with nature, governed according to customary laws and traditional knowledge systems that have ensured that they persist to the current day. There is a need to identify and understand how these conserved areas function, since they are patently effective and have persisted to the present day. Their custodians may not have stated explicit biodiversity outcomes, but because they hold values for these areas, they have addressed threats to their integrity through a wide variety of techniques, including taboos, local means to regulate use, nomadic lifestyles and so on. These 'other effective area-based conservation measures' do not need to be established, or have objectives set, but these need to be understood to better identify the wider range of tools available to society to conserve nature.

The studies analysed in this volume also reveal the hugely important social and economic outcomes that are intrinsic to biodiversity conservation, and the significant determining factors of governance and decision-making that can predict successful outcomes. A new generation of assessment techniques will place greater emphasis on the wider range of significant values that are embodied by protected and conserved areas, and the wider range of techniques beyond management effectiveness assessment itself that can elicit understanding of what leads to overall effectiveness. Included in these are the measures for the diversity, quality and equity of protected and conserved areas, to ensure the recognition of the identity and rights of custodians, to ensure upholding of systems for procedural justice including participatory approaches, and to ensure the equitable distribution of the costs and benefits of conservation. These 'means' to achieve effective conservation are also 'ends', in that they can create conditions for humanity living in harmony with nature.

Neglected by most assessment methods are those that must be applied at the system rather than the site scale. These include whether protected area systems properly represent all key biodiversity areas and the full diversity of ecosystem types. It concerns whether the systems themselves are ecologically connected and resilient, including through the restoration of corridors that span temperature and moisture gradients all for migration and movement in response to changing climate. It also concerns whether or not the full complement of governance authorities, including governments, private actors, indigenous peoples and local communities, can act collectively to ensure that the land use and marine spatial planning accommodate the need to conserve sufficient areas of terrestrial and inland waters, coastal and marine environments to allow for the persistence of biodiversity and its variety of values for humanity over time.

This volume will allow the reader to delve into these many variables at system and site scale, and help to identify those most efficient and effective ways to confront humanity's existential crisis in what has been called the Anthropocene, the period of humanity's history where one species has the power to destroy all others, and pose risks to its own future. Beyond scientific analysis and knowledge of what works in practice, it highlights and ponders the role of politics, economics and political will that hinder the application of the obvious remedies that can achieve sustainability for all.

Trevor Sandwith
Director, IUCN Global Protected Areas Programme,
Gland, Switzerland

Preface

2020 was intended to be the 'super year' for nature with the setting of a new Global Biodiversity Strategy and targets to achieve effective conservation of nature. The COVID-19 pandemic disrupted that timetable but not the ambition to change the trajectory of global conservation efforts and outcomes towards a more nature-positive future. The pandemic, with its likely origins in a spillover event from animals to humans that is a consequence of a degraded and fragmented natural world, reminds us that we depend on intact nature for our own well-being. Hopefully, we will see the successful conclusion of the post-2020 Global Biodiversity Strategy agreed in 2022 with ambitious new targets for the next decade and a positive plan to carry forward to 2050.

A key element of the strategy and a focus of targets and actions will be area-based conservation, primarily delivered through systems of protected areas and other effective area-based conservation measures (or OECMs in 'UN-speak'). A High Ambition Coalition for Nature and People (HAC) bringing together 70 countries is promoting ambitious targets for effective area-based conservation. IUCN and many international NGOs support this ambition and together with the HAC they are promoting a 30 × 30 target of protecting at least 30% of land and ocean by 2030. If achieved, this would see a significant expansion of the ***quantity*** of protected and conserved areas.

This book, *Effectiveness of Protected Areas in Conserving Biodiversity: A Worldwide Review* by David Rodríguez-Rodríguez and Javier Martínez-Vega, examines one of the key questions around area-based conservation – the question of effectiveness or ***quality***. It seems obvious that quantity of protected and conserved areas without quality will not produce the outcomes for nature and people that are sought. However, evidence tells us that quality is an issue that demands more attention if we are to achieve a nature-positive future.

Using a systematic review of available peer-reviewed literature, the authors have assembled available evidence. The first point to make is that we need to spend more time considering the quality of governance, management and outcomes of area-based conservation. They found only 76 studies that met their criteria for inclusion. Based on these studies, they seek to answer this question of how effective management is, and to tease out what we understand of the factors that influence effectiveness. They have done a service to the conservation world through this analysis of available data. It is not a simple issue to address, and they have added to a more nuanced understanding by consideration of case studies of both success and failure from around the world. This book will be of interest to academics who study protected and conserved areas but also to managers and others who work on the front line of conservation and who want to maximise the effectiveness of their management.

St. Lucia, QLD, Australia Marc Hockings

Preface

Protected and conserved areas constitute a key tool in the effort to conserve nature and halt the loss of biodiversity. And, as human pressure on our ecosystems continues to increase, leaving over 75 per cent of Earth's terrestrial surface significantly impacted by humans, and with more species than ever in human history at risk of extinction – the need for an effective global network of protected and conserved areas has never been more important. This is also recognized by the United Nation's Convention on Biological Diversity through which its 196 signature parties agreed to conserve at least 17 per cent of terrestrial and inland water, and 10 per cent of coastal and marine areas by 2020. A goal that was almost reached and now looks to be surpassed by the Convention on Biological Diversity's next strategic plan: the 'Post-2020 Biodiversity Framework' which aims to ensure that at least 30 per cent of land areas and sea areas are protected globally by 2030. However, while aerial targets are important and have proven to be powerful policy levers for many countries to commit land to the protection of biodiversity, the same aerial targets have also turned out to be insufficient in terms of ensuring the effectiveness of those protected and conserved areas. Thus, even with more than 270,000 individual sites registered in the World Database on Protected Areas, it is clear that the global network of protected and conserved areas has not sufficiently covered areas of particular importance for biodiversity nor is the network ecologically representative or well connected. Further, while ensuring protected and conserved areas are located where they best cover the immense diversity of life on Earth is of paramount importance – protected and conserved areas can only hope to conserve this biological diversity if they are effectively and equitably managed.

In addition, while protected and conserved areas are inherently a tool developed for nature conservation, a global target to protect 30 per cent of land and sea areas cannot be achieved if this adversely impacts the people living in and around them. This has led to an increasing awareness of the interplay between biodiversity conservation and the livelihood of the communities living in and around protected and conserved areas. This interplay has proven highly complex with both huge potentials for synergies as well as potentials for trade-offs where the conservation of biodiversity might negatively influence the opportunities and rights of people – people who are often already marginalized. Further, while protected and conserved areas now have the dual objective to safeguard biodiversity and contribute to the livelihood of people, those same people are also instrumental in the effective and equitable management of the same sites. In the words of Adams and McShane from their 1992 book *The myth of wild Africa*, 'Conservation will either contribute to solving the problems of the rural poor who live day to day with wild animals, or those animals will disappear'. The acknowledgement of the role and rights of people in the conservation of biodiversity and in relation to protected and conserved areas has also resulted in one of the most significant augmentations to the very concept of 'protected areas' through the inclusion of 'Other Effective Area-Based Conservation Measures' – or OECMs. OECMs while incredibly diverse have importantly provided a more official role for community managed areas, indigenous territories and other governance types not easily encompassed under the umbrella of 'protected areas'. Understanding how these can help to conserve biodiversity in their own right and support the network of protected areas is critical.

It is in this light that the field of protected and conserved areas has evolved immensely over the last decade – both in the policy and academic spheres. And, it is in that light, that a comprehensive and updated book on the topic is needed. David Rodríguez-Rodríguez and Javier Martínez-Vega do a great job of providing such an updated guide to understanding the effectiveness of protected areas in conserving biodiversity that will be of great use to practitioners, students and academics alike. Importantly, the authors manage to cover both marine and terrestrial aspects of protected and conserved areas. While the objectives on land and in the sea are inherently the same, traversing these two major ecological realms has proven difficult due to differences in the threats facing the systems, the geo-physical conditions, management traditions, and often lack of collaboration between academics working within the field. Through the seven chapters of the book, Rodríguez-Rodríguez and Martínez-Vega deliver a review of the current state of the art related to understanding the effectiveness of protected and conserved areas and provide the historical context for the field. Additionally, in the first two chapters, they eloquently set the scene of the current biodiversity crisis and how protected areas fit into the diverse toolbox that is biodiversity conservation. This important and insightful book is thus a very timely review of protected and conserved areas that provides many important insights into how this essential conservation tool can be better used to deliver real conservation impacts in the next decade.

København, Denmark Jonas Geldmann

Acknowledgements

We would like to acknowledge each and every author of the studies that were included in this sistematic literature review. We would equally like to thank the authors and editors of the articles referenced in each case study for providing graphic material to illustrate the protected areas studied and their respective control zones. We are also thankful to the three worldwide-reputed protected area experts who kindly agreed to write some words to introduce this work to the readers. Without all of them, this book would have not been possible.Last but not least, we are also deeply grateful to Mr. Zachary Romano and his team at Springer for giving us the opportunity to publish this book.

Contents

Chapter 1
Human Impact on the Biosphere: A Contemporary Ecocide

Biodiversity is the diversity of life on Earth. Until now, biodiversity makes the single most distinguishing factor of planet Earth: its capacity to harbour life. The main international agreement to conserve biodiversity, the Convention on Biological Diversity, defines it as the 'variability among living organisms from all sources including, inter alia, terrestrial, marine and other aquatic ecosystems and the ecological complexes of which they are part; this includes diversity within species, between species and of ecosystems' (CBD 1992). This common definition implies three main levels of biological diversity, intra-specific (genetic diversity), specific (species diversity) and inter-specific (ecosystem diversity), with species being the common nexus among them. Species are commonly defined as groups of populations that can interbreed but cannot successfully reproduce with other groups producing fertile descendents. Therefore, reproductive success among populations is the barrier that clearly differentiates species among sexually reproducing organisms (Mayr 1942). Populations are the main evolutionary unit, whereas species are the main conservation unit (Woodruff 2001), although the conservation importance of populations is increasingly recognised (Coates et al. 2018).

Species range from the simplest single-cellular, microscopic beings to large, complex organisms such as elephants or humans. They are usually classified in five biological kingdoms (Margulis and Chapman 2009): Bacteria (single-cell or multicell prokaryotes), Protista (single-cell eukaryotes), Fungi (yeasts, rusts, smuts, mildews, molds and mushrooms), Plantae (plants) and Animalia (animals). Living beings aggregate in different levels of increasing complexity and spatial scale. Individuals of the same species (organisms) make a *population*. Interbreeding populations form a *species*. Populations of different species that dwell and interact in the same area and time define a *community*. The interaction of a set of species (community) and their physical environment defines a *habitat* or *ecosystem* (e.g. a pine forest). A set of similar ecosystems define a *biome* (e.g. taiga, or boreal forest) (Fig. 1.1).

D. Rodríguez-Rodríguez, J. Martínez-Vega, *Effectiveness of Protected Areas in Conserving Biodiversity*, Strategies for Sustainability,
https://doi.org/10.1007/978-3-030-94297-7_1

Fig. 1.1 Population of Asian elephants in their natural environment. (*Credit*: Javier Martínez)

Species adapt and span across many different environments in the biosphere: from deep oceans to vast plains through dark caves or even sulphuric eruptions. The total number of species known to science today is in the range of 1.8–2 million, with new species being discovered regularly. However, that diversity is likely just a small fraction of all existing life on Earth that is still awaiting 'discovery' (Purvis and Hector 2000; Ceballos et al. 2010). Terrestrial species are far more diverse than marine ones, especially in the tropics (Costello 2015). Yet it is estimated that more than 90% of marine animal species are still unknown to science (Phillipe et al. 2011; Just et al. 2014). Lack of species knowledge makes also a large gap in rich tropical areas (Young et al. 2016).

That extraordinary diversity of life in the form of genes, species and ecosystems, so far unique in the entire universe, is called 'biodiversity'. Biodiversity results from a multi-million-year evolutionary process whereby individuals randomly acquire new features that allow them to better (or worse) adapt to their environment as a result of accidental genetic changes called 'mutations'. If those genetic changes by some individuals are broad enough and maintained in time, that is, transmitted to their descendents, they may lead to different subspecies (genetically distinct populations still within the same original species) or even to new species from the one those individuals initially belonged to in a process called 'speciation'. Evolution and its long-term consequence, speciation, is normally a lengthy, gradual process spanning over millennia, although rapid genetic change leading to different subspecies or even species may occur in the course of few generations, especially among short-lived organisms. Evolution may lead to greater biodiversity by increasing the diversity of genes, species and ecosystems, although it may also lead to extinction, if the random genetic changes affecting individuals and then populations negatively influence their ability to adapt to their environment. Thus, most species that ever lived on Earth became extinct as a result of their insufficient ability to adapt to their changing environments. This may have occurred as a result of rapid, acute environmental changes or through slower evolutionary processes such as intense predation, disease and competition with other abler species. For instance, one of the most popular theories explaining the mass extinction of dinosaurs in the late Cretaceous was a drastic change in atmospheric composition resulting from the impact of a large asteroid off the coast of today's Eastern Mexico.

The Earth has experienced five mass extinctions in which large numbers of existing species became extinct in a relatively short period in geological terms. All these mass extinctions share some common features: they occurred in relatively short time periods, geologically speaking, spanning from thousands of years to millions of years; they resulted in the eradication of large proportions of living species; their impact was not random taxonomically, thus affecting some groups of species over others; the survivors were often not previously dominant groups; and they were driven by broad geological events (Ceballos et al. 2010; Table 1.1). After such dramatic events, the remaining life forms diversified at different rates as a result of new ecological opportunities from the many empty niches left by extinct species. In such broad evolutionary events, recovering biodiversity values prior to mass extinction episodes took millions of years. It is thought that the current number of living species, ranging between 3 million and 30 million, are probably just under 0.1% of

Table 1.1 Main characteristics of the six mass extinctions on Earth

Mass extinction	Geological era	Date	Likely driver	Impact on biodiversity (% species lost)
1st	Ordovician-Silurian extinction	444 million years ago	Global cooling from orogenic movements	85
2nd	Late Devonian extinction	383-359 million years ago	Atmospheric change from volcanism	75
3rd	Permian-Triassic extinction	252 million years ago	Global warming from volcanism	75 on land and 96 at sea
4th	Triassic-Jurassic extinction	201 million years ago	Global warming from volcanism	80
5th	Cretaceous-Paleogene extinction	66 million years ago	Global cooling from asteroid impact	76
6th	Anthropocene	Twentieth and twenty-first centuries AC	Human activity	?

Source: https://www.nationalgeographic.com/science/prehistoric-world/mass-extinction/

Fig. 1.2 Triceratops' skeleton. Natural History Museum, London. (*Credit*: David Rodríguez)

all living organisms that ever lived on Earth (Margulis and Chapman 2009) (Fig. 1.2).

Though the current amount of biodiversity is thought to be the largest in the history of Earth (Ceballos et al. 2010), many authors now recognise that we have entered a new mass extinction event, the sixth one (Ceballos et al. 2015). Similar to the previous ones, it is resulting in large rates of biodiversity loss, estimated at between 11,000 and 58,000 extinct species annually (Dirzo et al. 2014). However, the ongoing mass extinction has two distinct features never seen before: it is occurring at an alarming speed; and it is driven by a single species: humans.

The human species is a singular one. Being just one of the approximately 2 million species known to science, its impact on the biosphere is immense, even greater than other highly biosphere-impacting species such as phytoplankton. The human species alone appropriates between 25% and 40% of all net primary productivity on Earth, consumes 35% of the productivity of the oceanic shelf, uses over 60% of freshwater run-off, occupies 40% of land and directly impacts 83% of the Earth's surface and 87% of the oceans (Sanderson et al. 2002; Krausmann et al. 2013; Jones et al. 2018). Common human activities such as agriculture, industry, farming, fishing, mass consumption, hunting and tourism are driving major pressures on global biodiversity such as ecosystem degradation, overexploitation of wild populations, introduction of alien species, pollution of habitats, or climate change. Of those pressures, ecosystem degradation and destruction outstands as a major factor of biodiversity decline, affecting up to 85% of threatened species (Pereira et al. 2012; Young et al. 2016). Some figures provide clear evidence of the impact of human-driven land use changes on the Earth's biodiversity: up to 85% of all ever-existing wetland area has been lost (IPBES 2019); deforestation due to logging, agricultural expansion and mining is a large-scale issue in the biodiversity-rich tropics, with an estimated net loss of 7% between 2000 and 2013 (Potapov et al. 2017). As a result, global tree-covered area is estimated at just over 54% of its prehuman value, with only 39% of all land area covered by primary, non-altered vegetation (IPBES 2019) (Fig. 1.3).

Humans started to widely transform natural habitats with the Neolithic spread of agriculture some 10,000 years ago. Low-density, nomadic Paleolithic humans are thought to have had relatively low environmental impact through hunting and gathering activities, although they are likely to have contributed to the extinction of large charismatic megafauna, such as mammoths (Johnson et al. 2017). Domestication of plants and animals allowed human populations to settle down, store food surpluses and grow in numbers. For that, they needed space, so they made it. Forests, savannahs and shrublands were cut down or burned; rivers and wetlands were diverted, channelled or desiccated to give way to crops and grasslands for farm animals. They also needed to protect their food supplies, so they started persecuting competing or dangerous fauna to them, their crops or animals. Since those days, the human species has become a major ecological driver on the Planet. Some authors even deem that extraordinary evolutionary feature a new geological era of the Earth's history:

Fig. 1.3 Agriculture is the main land use globally. (*Credit*: David Rodríguez)

the Anthropocene (Lyons et al. 2016). Other authors prefer the pre-industrial baseline of 1500 as the beginning of the new era when the human species undisputedly dominated the Earth's ecological processes (Dirzo et al. 2014). In any case, all of them agree that humans have long exceeded the regeneration capacity of the Planet and crossed dangerous ecological limits, especially since the Industrial Revolution when environmental degradation increased rapidly (MEA 2005; Steffen et al. 2015). Actually, as much as half of the total human-induced environmental change in the history of humankind has been done since the mid-twentieth century, with accelerated population growth and mass consumption and the ultimate drivers of environmental degradation (IPBES 2019).

Human-made pressures are rapidly and seriously depleting our Planet of life (WWF 2020). Conservative figures on recent animal species' extinctions report 617 vertebrate species extinct since 1500, 477 of them since 1900, with a clearly accelerating pace since the Industrial Revolution (Ceballos et al. 2015). Such unfortunate species would have still naturally lived among us for between 800 and 10,000 years more, had the humans not speeded up their extinction (Ceballos et al. 2015). Those are conservative figures on the small fraction of existing species *that have been described to science and whose extinctions have been documented*. It is most likely that real human-driven extinction figures are substantially higher, as numerous species are thought to become extinct recurrently before being even described by science (Ceballos et al. 2010). Yet, the number of species that are threatened with extinction is much higher than documented historical extinctions, with 1665 such species reported by the IUCN in a sample of 32,765 species in 2008 (excluding those surviving in captivity) and a global estimate of nearly 258,000 species threatened with extinction (Ceballos et al. 2010), roughly 14% of all the species known to science at that date. More recent estimates increase that figure to 25% and a global loss of up to 1,000,000 species within decades if current pressures continue unabated (IPBES 2019).

Whereas natural, background extinction rates driven by evolutionary processes have been estimated at between one and two species per million of species per year, current human-driven rates are well over 100 times the background rate, at least for the past 500 years (Dias et al. 2017; Gaffney and Steffen 2017). Some predictors of faunal extinction risk include small geographic range, low reproductive rate, large home range size and large body size (Dirzo et al. 2014).

But not only are plenty of species becoming extinct as a result of human activities. The number of populations and individuals of the remaining species is also declining in a process called 'defaunation' for animals and 'deforestation' for plants (Dirzo et al. 2014). Defaunation is more intense in biodiversity-rich tropical regions on land and, above all, in freshwater ecosystems (WWF 2020), with amphibians as the most affected vertebrate taxa, with 41.5% of species threatened with extinction (Young et al. 2016). Invertebrate species are thought to experience similar declining trends to vertebrates, even though they have been scarcely studied. Of those invertebrate species that have been assessed, 40% are threatened with extinction, and many other species' populations are decreasing including grasshoppers, beetles, dragonflies, ants, wild bees and butterflies (Dirzo et al. 2014). Among plants, cycads

Fig. 1.4 Amphibians are the most endangered vertebrate taxa. (*Credit*: David Rodríguez)

and gymnosperms face similar or greater risks of extinction than most vertebrate groups, with over 63% and 40% of threatened species, respectively (IPBES 2019) (Fig. 1.4).

Effects of species' extinction include loss of biological diversity; shifts in abundance, composition and behaviour of related species; homogenisation of biotas; degradation of ecosystem functions; cascade extinctions; disruption of evolutionary processes; individual downsizing; and changes of ecosystem services (Dirzo et al. 2014; Young et al. 2016). Whereas those are some 'evident' effects of species' extinction, belated, cryptic extinctions are likely to occur after habitat degradation has reached certain limits in what is called 'extinction debt' (Tilman et al. 1994). A minimal amount of intact habitat is needed to sustain thriving animal populations, but some habitats, such as mangroves, seagrass meadows and other marine habitats, have already gone under that limit globally (IPBES 2019), and most remaining ecosystems are becoming increasingly smaller and fragmented (Jacobson et al. 2019), being thus unfit to sustain human-sensitive, specialist and broad ranging species in the long term (Pullin 2002). Around 500,000 species are conservatively thought to be bound for extinction for dwelling in insufficiently large or poor quality habitats, including over 3000 vertebrates and 40,000 plants (IPBES 2019).

Humans are not only reducing the amount of biodiversity on Earth. We are also reducing its quality. As a result of wildlife extinctions, species introductions and farming, biotas are becoming ever more similar (IPBES 2019). We can progressively observe the same few species everywhere we go. Actually, some dozen human-bred or cultivated species such as wheat, maize, chicken or cows are thriving in increasingly larger parts of the Earth, with livestock now accounting for over 90% of megafaunal biomass on land (Barnosky 2008). Human-made pressures, namely hunting, fishing, land use changes and climate change, are successfully fighting against evolution and eroding wild and domestic populations' genetic variability alike, with an estimated 1% loss per decade in genetic diversity since the mid-nineteenth century (Leigh et al. 2018). Countless genetically distinct wild populations and domesticated breeds have gone extinct, and many others face the same prospect (IPBES 2019). Selective extraction of individuals with exceptional traits such as long horns or large body sizes through trophy hunting or fishing is similarly working against natural selection and leading to a modest, less biologically fit set of remaining populations.

By degrading biodiversity, we are creating an increasingly poor, homogeneous and vulnerable Planet, and increasing the risk of further extinctions, including ours. Less biodiversity not only entails less provision of ecosystem services, of human well-being, but also less capacity to recover from environmental stresses. This large-scale eradication of life on the Planet has been deemed an 'ecological genocide' or 'ecocide' (Kalkandelen and O'Byrne 2017; Crook et al. 2018). 'Genocide' implies active and planned intention to eradicate a people, race, ethnicity or other human group (United Nations 2021). It is hardly arguable that humans are eliminating most species with a planned and mindful intention to do so. Deliberate human-driven extinction of some species, mostly predators such as the Tasmanian wolf, has occurred to protect human interests (Fig. 1.5). However, conscious eradication of

Fig. 1.5 Dissected Tasmanian wolf in a museum. (*Source*: National Museum of Natural History, Madrid https://es.wikipedia.org/wiki/Thylacinus_cynocephalus#/media/Archivo:TilacinoMadrid.jpg)

living species is more the exception than the rule in human-made extinctions. Rather, it seems clear that humans are pushing plenty of other species to extinction as a side effect of a resource-intensive development model that is almost entirely decoupled from nature (Azqueta et al. 2007). Thus, in our view the concept of 'genocide' does not generally and adequately apply to the modern human-nature relationship.

Some scholars claim that impacts to human communities as a result of environmental degradation need to be considered and incorporated to the legally backed concept of genocide, thus providing recognition to the destruction of other living beings, even if indirectly, through the concept of 'ecocide' (Kalkandelen and O'Byrne 2017; Crook et al. 2018). We think that the intentional destruction of biodiversity should be considered an international crime *regardless of its impacts on humans*. A recently agreed international definition of 'ecocide' aimed at amending the Rome Statute on the International Criminal Court addresses our concerns more appropriately: '…"ecocide" means unlawful or wanton acts committed with knowledge that there is a substantial likelihood of severe and either widespread or long-term damage to the environment being caused by those acts', 'Environment' meaning the earth, its biosphere, cryosphere, lithosphere, hydrosphere and atmosphere, as well as outer space (Stop Ecocide Foundation 2021). According to this definition, much of our contemporary and current relationships with nature could be considered cases of ecocide.

There are interesting philosophical, moral and legal considerations regarding the 'intrinsic value' of other species, their right to exist and the right of humans to make any use of biodiversity at will. However, these are beyond the scope of this book. Other more pragmatic views in favour of conserving biodiversity come from its value to humans, whether as direct value (as raw materials, for instance) or option value (the possibility of using biodiversity in the future). This 'utilitarian' view to biodiversity conservation is the one followed by the 'ecosystem service' approach to conservation, now mainstream in conservation science and policy (MEA 2005; IPBES 2019).

Ecosystem services are defined as 'the conditions and processes through which natural ecosystems, and the species that make them up, sustain, and fulfill human life' (Daily 1997). Or, put more simply: 'the benefits people derive from ecosystems' (MEA 2005). Three groups of ecosystem services are broadly acknowledged according to the type of service that supporting biodiversity provides to humans: provisioning, regulating and cultural (TEEB 2010). Provisioning ecosystem services include food, fibre, genetic resources and drinking water. Regulating services relate to vital ecosystem functions such as air quality regulation, climate regulation, pollination and pest control. Cultural services include spiritual, aesthetic and recreational services. Whereas few provisioning services such as crop or livestock production have increased in recent years as a result of intensive farming practices, most other services that we humans enjoy have decreased as a result of the degradation of supporting biodiversity (MEA 2005). That means less species and also less well-being. In the short term, biodiversity contributes to the adequate ecosystem's functioning (Dirzo et al. 2014). In the long term, it increases ecosystems' resistance and resilience to environmental pressures and stresses (Tinaut and Ruano 2002) (Fig. 1.6).

Fig. 1.6 Water provision and water quality regulation are two vital services provided by healthy ecosystems. (*Credit*: David Rodríguez)

Some studies have put a monetary price to all the services that biodiversity provides to humans for free and estimated a benefit range between 125 and 145 trillion USD (in 2007) per year, with 4.3–20.2 trillion USD (in 2007) lost every year as a result of ecosystem degradation between 1997 and 2011 (Costanza et al. 2014). Yet numerous studies concur that ecosystem services and underlying biodiversity continue to be degraded at an unprecedented rate and scale, severely impacting many species and human well-being (MEA 2005; IPBES 2019). It is widely advocated that the current biodiversity crisis will not be averted until the two main drivers of biodiversity loss are properly addressed: excessive human population and individual consumption of goods and services. Utopian as it may seem to regulate births and consumption in an increasingly liberal and capitalist world, a range of less ambitious initiatives to reduce biodiversity loss has been devised and implemented. One of them, to which this monograph is dedicated, is setting aside places especially devoted to the conservation of biodiversity where impacting human activities are restricted or forbidden in order to let other species survive.

Chapter 2
Strategies to Address the Biodiversity Crisis

The huge and mostly harmful impact of humans on the biosphere was internationally acknowledged at the UN Conference on the Environment in Stockholm in 1972. Some previous initiatives had addressed conservation concerns of few species or individual ecosystems such as wetlands. The Stockholm Conference was the first international meeting that made the environment a major issue. Two major milestones arose from that event: a theoretical one, the worldwide known and widely used concept of 'sustainable development', and a practical one, the creation of the United Nations Environment Programme (UNEP). Since those days, a number of international agreements have tried to address the global loss of biodiversity (Table 2.1).

In addition to those shown in Table 2.1, other biodiversity-relevant regional or global agreements include the Barcelona Convention for the protection of the Mediterranean Sea (1976), the Abidjan Convention for the protection and development of the marine environment of the west and central African region (1981), the Lima Convention for the protection of the south-east Pacific (1981), the Jeddah Convention for the Conservation of the Red Sea and the Gulf of Aden (1982), the Cartagena Convention for the protection of the Caribbean Sea (1983), the Nairobi Convention for the protection of the marine environment of the eastern African region (1985), the Noumea Convention for the protection of the south Pacific (1986), the Framework Convention on Climate Change (1992), the OSPAR Convention for the protection of the marine environment of the north-east Atlantic (1992), the HELCOM Convention on the protection of the Baltic Sea (1992), or the Framework Convention for the protection of the Caspian Sea (2003).

The above-mentioned agreements contain several commitments and recommendations aimed at different stakeholders, chiefly national governments, at different spatial scales. Such commitments and recommendations can be broadly divided into two complementary strategies for biodiversity conservation: (1) *ex situ* conservation and (2) *in situ* conservation.

D. Rodríguez-Rodríguez, J. Martínez-Vega, *Effectiveness of Protected Areas in Conserving Biodiversity*, Strategies for Sustainability,
https://doi.org/10.1007/978-3-030-94297-7_2

Table 2.1 Most relevant international agreements for the conservation of the world's biodiversity

Agreement	Leading body	Signature date	Objectives	Geographic scope	URL
International Convention for the Regulation of Whaling	Government of the United States of America	1946	Establish the International Whaling Commission; scientific research on whales and whaling; conservation and sustainable use of whales	World	https://iwc.int/convention
Convention on Wetlands of International Importance Especially as Waterfowl Habitat	International Waterfowl and Wetlands Research Bureau	1971	Conservation and wise use of wetlands and waterfowl	World	https://www.ramsar.org/sites/default/files/documents/library/current_convention_text_e.pdf
Convention Concerning the Protection of the World's Cultural and Natural Heritage	UNESCO	1972	Protection, conservation and presentation of the cultural and natural heritage	World	https://whc.unesco.org/en/conventiontext/
Convention on International Trade in Endangered Species of Wild Fauna and Flora	IUCN	1973	Ensure that international trade in specimens of wild animals and plants does not threaten their survival	World	https://cites.org/eng/disc/text.php
Convention on the Conservation of European Wildlife and Natural Habitats	Council of Europe	1979	Conservation of wild flora and fauna species and their habitats	Europe	https://www.coe.int/en/web/bern-convention/presentation
Convention on the Conservation of Migratory Species of Wild Animals	United Nations	1979	Conservation and sustainable use of migratory animal species and their habitats	World	https://www.cms.int/en/convention-text
Protocol on Environmental Protection to the Antarctic Treaty	Secretariat of the Antarctic Treaty	1991	Comprehensive protection of the Antarctic environment and dependent and associated ecosystems; designation of Antarctica as a natural reserve, devoted to peace and science	Antarctica	https://www.ats.aq/e/protocol.html

Agreement	Leading body	Signature date	Objectives	Geographic scope	URL
Convention on Biological Diversity	United Nations	1992	Conservation of biological diversity; sustainable use of its components; fair and equitable sharing of benefits arising from genetic resources	World	https://www.cbd.int/convention/text/
Agreement on the Conservation of Small Cetaceans of the Baltic, North East Atlantic, Irish and North Seas	Secretariat of ASCOBANS	1992	Achieve and maintain a favourable conservation status for small cetaceans	Europe	https://www.ascobans.org/en/documents/agreement-text
United Nations Convention to Combat Desertification	United Nations	1994	Combat desertification and mitigate the effects of drought in countries experiencing serious drought and/or desertification	World (Africa)	https://www.unccd.int/convention/about-convention
Agreement on the Conservation of Cetaceans of the Black Sea, Mediterranean Sea and Contiguous Atlantic Area	Secretariat of ACCOBAMS	1996	Achieve and maintain a favourable conservation status for cetaceans	Black Sea, Mediterranean Sea and Eastern Atlantic	https://www.accobams.org/wp-content/uploads/2017/01/ACCOBAMS_Text_Agreement_English.pdf
International Treaty on Plant Genetic Resources for Food and Agriculture	FAO-United Nations	2001	Conservation and sustainable use of plant genetic resources for food and agriculture; fair and equitable sharing of the benefits arising out of their use	World	http://www.fao.org/3/i0510e/i0510e.pdf
Agreement on the Conservation of Albatrosses and Petrels	ACAP Secretariat	2001	Achieve and maintain a favourable conservation status of albatrosses and petrels	World	https://www.acap.aq/agreement-text/206-agreement-on-the-conservation-of-albatrosses-and-petrels/file

Fig. 2.1 A number of international agreements have addressed marine conservation issues. (*Credit*: Francisco Jiménez)

Ex situ conservation entails the conservation of selected plant and animal species in facilities outside their natural habitats, such as botanic gardens, zoos, aquariums or gene banks. Gene banks are aimed at preserving genetic diversity in the form of reproductive cells or seeds of wild and domestic species under carefully controlled conditions to prevent extinction of whole species or varieties of human use, cultural value or conservation interest. Perhaps the most famous gene bank is the Svalbard Global Seed Vault in Norway (Svalbard Global Seed Vault 2021; Westengen et al. 2013). It was created following the entry into force of the International Treaty on Plant Genetic Resources for Food and Agriculture in 2004 and aims at preserving viable samples of all edible plant species of the world.

Some groups in industrialised societies oppose the maintenance of wildlife in captivity on animal welfare grounds. Whereas wild animals should certainly be rightful to live in their natural habitats, when human activities make it impossible or too risky for them to dwell in their historic domains due to insufficient remaining habitat or poaching, captivity becomes their only chance for survival. Actually, the IUCN accounted for this fact and produced a threatened species' Red List category of 'Extinct in the wild' meaning that such species only survives in captivity (IUCN 2012). For those extremely valuable species on the verge of extinction, human care in *ex situ* facilities allows them to hang on to existence.

Botanic gardens and zoos play two essential roles for biodiversity conservation: environmental awareness and captive breeding. It is estimated that over 700 million

Fig. 2.2 Andean condors in captivity. (*Credit*: David Rodríguez)

people visit zoos and aquariums annually, potentially making an enormous impact on environmental education and awareness globally (Gusset and Dick 2011).

Around 25% of all described bird species, 20% of mammals, 12% of reptile species and 4% of amphibians are currently kept at zoos (Conde et al. 2011). This is a very substantial amount of the world's biodiversity, even more so if we consider the very limited area that such facilities cover. Human populations are increasingly urban and detached from their natural environment. Thus, many people, especially children and teenagers, know little about the natural world and usually have not much chance to experience it by themselves. Therefore, botanic gardens and zoos are sometimes their only way of getting acquainted with wild species that live very far away and are extremely difficult to find in their natural habitats. It is known that care and appreciation comes with knowledge (Clayton and Myers 2015). Thus, visits to such facilities whether autonomously or enrolled in organised educational activities are very likely to increase people's valuation of such species as well as their awareness of their situation, making visitors more prone to pro-environmental behaviour that will probably benefit those species either directly or indirectly (Arendt and Matthes 2014; Bruni et al. 2008).

The other vital role of botanic gardens and zoos is reproduction of endangered species. Captive breeding aims at increasing the population size of endangered species to reduce their immediate risk of extinction and, in the medium term and if conditions are favourable, planning the reintroduction of some individuals in the wild. Zoological networks work together in breeding programmes of endangered species, exchanging expertise and individuals to maximise reproductive success and avoid genetic degeneration through inbreeding. One example of such international effort is the European Endangered Species Programme of the European Association of Zoos and Aquaria (EAZA 2021). The World Association of Zoos and Aquarium (WAZA)'s members are spending nearly US$ 350 million per year on conservation projects in the wild, making WAZA one of the largest contributors to biodiversity conservation in the world (Conde et al. 2011). Despite the fact that only about one in ten reintroduction actions have succeeded, such networks' efforts have produced inspiring outcomes for a number of species that have been reintroduced successfully in their natural habitats where they now maintain viable wild populations. The European bison, the Przewalski's horse, the Iberian lynx, the Partula snail and the Californian condor are examples of the importance of captivity breeding programmes, even more so as human interventions continue to degrade natural habitats and populations. As a result, it is becoming increasingly common that larger proportions of entire species' populations are kept in captivity (Conde et al. 2011). Similarly, botanic gardens harbour at least 105,634 species, equating to 30% of all plant species diversity, and conserve over 41% of known threatened species (Mounce et al. 2017). On the one hand, these data show that biodiversity degradation continues unabated and that human activities keep pushing plenty of species over the edge in their natural habitats. On the other hand, *ex situ* conservation tools mean a safeguard that a relevant number of endangered species could eventually be saved from extinction.

The other generic conservation strategy is *in situ* conservation. The CBD defines *in situ* conservation as 'the conservation of ecosystems and natural habitats and the maintenance and recovery of viable populations of species in their natural surroundings' (CBD 1992). The most widely used on-site conservation tool is protected areas (PAs). The IUCN defines PAs as 'clearly defined geographical spaces, recognised, dedicated and managed, through legal or other effective means, to achieve the long term conservation of nature with associated ecosystem services and cultural values' (Dudley 2008). PAs will be discussed in more detail in the next chapter.

Other effective area-based conservation measures (OECMs) are another recent (CBD 2010), closely related on-site conservation tool. The CBD provides a very similar definition of OECMs to that of PAs: 'a geographically defined area other than a Protected Area, which is governed and managed in ways that achieve positive and sustained long-term outcomes for the in-situ conservation of biodiversity, with associated ecosystem functions and services and, where applicable, cultural, spiritual, socio-economic, and other locally relevant values' (CBD 2018). As it can be easily spotted, there is just a methodological and a conceptual difference of OECMs with regard to PAs: OECMs cannot be PAs at the same time, and they do not need to be intended at biodiversity conservation as their first goal. Instead, conservation can be achieved as an ancillary outcome of the site's management or designation for whatever reason. An example of such sites might be zones around offshore wind farms, which create exclusion zones to navigation, mineral exploitation, fishing and other activities, which are likely to benefit biodiversity in the area.

Fig. 2.3 Botanic gardens play an important role in plant conservation. Kew Gardens, London. (*Credit*: David Rodríguez)

The fact that an official definition of OECMs was not adopted until 2018 has most likely deterred scientists and decision-makers from making wider use of this tool. By August of 2021, only three countries had officially recognised OECMs in their territories and reported them internationally: Canada, Algeria and South Africa (Protected Planet 2021). OECMs are, however, envisaged to play an important role in achieving effective conservation (IUCN-WCPA 2019) and helping countries to meet oncoming global biodiversity targets for 2030 (CBD 2015, 2016).

Another important *in situ* conservation tool is ecological restoration. Ecological restoration is the process of assisting the recovery of an ecosystem that has been degraded, damaged or destroyed (SER 2021). Therefore, its primary focus is not preventing species' extinction and ecosystems' degradation but reverting them. A number of restoration approaches that may entail specific ecological improvements or full ecosystem's reversion can be followed, depending on the restoration

Fig. 2.4 Restored ecosystems become increasingly similar to the original ecosystem as time passes. Restored oak tree forest, Portugal. (*Credit*: David Rodríguez)

objectives. Such approaches range from mere species' population reinforcements through individuals' translocation or release from captive breeding to whole ecosystems' recreation. Well-planned, sufficiently funded ecological restoration actions allow bringing even extremely degraded ecosystems (e.g. built-up areas) close to their original, pristine conditions.

If the restoration objectives dictate so or if the environmental conditions of the degraded site have greatly changed from their original conditions, making it impossible to restore it to its previous state, ecosystem's replacement or recreation can also be brought about. Such restoration approaches may produce ecologically functional ecosystems that did not exist in the area before. An example is opencast mining. Opencast mining drastically changes the environmental characteristics of the affected area by removing large quantities of material and creating bare rock ground and holes. Such holes could be used, if conditions allow for it, to recreate valuable wetland habitats by digging up to the groundwater level and letting groundwater emerge and cover the previously sterile rocky ground. Therefore, a previous hilly forested area may become an ecologically valuable wetland through ecological restoration.

From a technical point of view, two general ecological restoration strategies can be followed: passive restoration and active restoration (Trujillo-Miranda et al. 2018). They differ in the intensity, cost and time required to achieve the required objectives. Passive restoration entails the prevention or removal of the causes of degradation of the species or ecosystems of concern in order to let biodiversity recover by itself. For instance, a wild population endangered by poaching may be passively restored through managerial actions that eliminate poaching without actual intervention on the species. Once poaching is eradicated from the area, the species may recover safe numbers in some years following natural reproductive success. Sometimes, passive restoration involves doing little or even nothing at all. For example, flat pine forest areas may entirely recover their previous state from wildfires without any human intervention, if enough time (often decades) passes without further natural or human-made interference. Passive restoration requires limited effort and cost, but takes a long time to render the desired outcomes.

In contrast, active restoration usually implies high intensity efforts in terms of machinery, human resources and funding to restore ecosystems or wild populations. If the previous burnt pine forest was not in a flat area but in a mountain area in a Mediterranean region, some mechanical soil-protection actions such as the creation of soil barriers and terraces, seeding and planting new trees will probably be required to prevent soil loss through wind erosion and torrential rain. Such intensive intervention often produces rapid results but also comes at a high price, which is not always possible to pay in many poor countries and regions.

Focusing on expanding suitable spaces for biodiversity, ecological restoration is a powerful tool to improve the status of biodiversity. As such, the UN has declared 2021–2030 the Decade on Ecosystem Restoration, aimed at preventing, halting and reversing the degradation of ecosystems on global lands and at seas (UNEP-FAO 2021). Nevertheless, we should take some important considerations into account before thinking of ecological restoration as a panacea to the global loss of biodiversity. Firstly, ecological restoration, even if well planned, done and funded, is unlikely to fully recover the ecological characteristics of original ecosystems. It may restore most of the sites' initial biodiversity and ecosystems' functions, but completely restoring ecological processes is often extremely difficult, sometimes impossible (Almassi 2017; Keulartz 2012). Secondly, even if restoration actions are carefully planned and adequately implemented, each ecosystem is a complex, multifactorial entity with several interactions that may not always be possible to consider. Thus, success in ecological restoration is rarely guaranteed, and a large number of restoration projects do not fully achieve their intended objectives, or do not achieve them at all (Herrick et al. 2006; Palmer et al. 2010). Thirdly, ecological restoration, especially active restoration, is expensive, and available resources for biodiversity conservation are very often limited, even in highly developed, rich countries. That means that large proportions of degraded habitats and species will count on no funding or little funding to be restored (Carabias et al. 2007; Ceccon et al. 2015). Lastly, poorly planned or executed restoration actions may entail substantial ecological damage, for instance through the use of exotic species in the restoration process, which may later become invasive and a threat to native biodiversity (Wang et al. 2013).

These considerations suggest that, although undoubtedly useful to revert the degradation trends of global biodiversity, ecological restoration is probably playing an ancillary, spatially limited role in biodiversity conservation with regard to other *in situ* conservation tools aimed at conserving genes, species and ecologically functioning ecosystems prior to their degradation, such as PAs or OECMs. Conserving original habitats is always ecologically more effective than restoring them, and it is also often cheaper.

Chapter 3
Protected Areas

PAs are the main, most widely used conservation tool to address the loss of biodiversity globally. Most environmental protection agreements include PAs as a possible or mandatory measure to enhance the status of biodiversity. In 2010, the CBD's Conference of the Parties approved the Strategic Plan for Biodiversity 2011–2020 and agreed on global conservation targets, including enlarged PA coverage on land (17%) and sea (10%) that should be met by 2020: the Aichi Biodiversity Targets (CBD 2010). As a result, PAs currently cover 15.7% of global terrestrial and freshwater ecosystems and 7.7% of marine ecosystems, a relevant international achievement though falling short of the agreed targets by 2020 (Protected Planet 2021).

There are two internationally agreed definitions of 'PA'. Article 2 of the CBD defines PA as 'a geographically defined area, which is designated or regulated and managed to achieve specific conservation objectives' (CBD 1992). The more recent definition by the IUCN defines PAs as 'a clearly defined geographical space, recognised, dedicated and managed, through legal or other effective means, to achieve the long term conservation of nature with associated ecosystem services and cultural values' (Dudley 2008). Both definitions coincide in PAs' basic elements: *in situ*, area-based legal and managerial measures established to conserve biodiversity. From a non-utilitarian point of view on biodiversity, PAs are a form of self-restriction to development and human well-fare and as such can be regarded as altruistic tools. However, one could argue that even legally stringent 'reserves' serve some human interests, even if compatible with other species' right to exist, such as science production, education or tourism.

PAs are not a new conservation tool. The first modern PAs were designated in the late nineteenth century in the United States of America largely on scenic grounds: Yellowstone and Yosemite National Parks. Today, virtually all countries and territories have designated PAs in their domains with diverse objectives. However, until recently PAs have been chiefly established in economically unproductive, remote

D. Rodríguez-Rodríguez, J. Martínez-Vega, *Effectiveness of Protected Areas in Conserving Biodiversity*, Strategies for Sustainability,
https://doi.org/10.1007/978-3-030-94297-7_3

Fig. 3.1 Ranomafana National Park, Madagascar. (*Credit*: David Rodríguez)

Fig. 3.2 Protected areas sustain most remaining populations of large carnivores. (*Credit*: David Rodríguez)

places where they seldom interfere with human activities (Gaston et al. 2008; de Oliveira Soares and Costa Lucas 2018). Consequently, many PAs are not located in the most important areas for biodiversity. Actually, only 19% of all global Key Biodiversity Areas (KBAs) are fully included in PAs, whereas 39% of them are completely unprotected (Protected Planet 2020). KBAs are sites that contribute significantly to the global persistence of biodiversity for a number of reasons, including hosting a substantial proportion of the total population size of threatened species covering a substantial proportion of the total extent of threatened ecosystems, hosting a substantial proportion of the total population size of species (or total extent of ecosystems) with restricted geographic ranges, harbouring ecologically intact communities, and site's biological irreplaceability (IUCN 2016).

National park, protected forest, scenic landscape, nature monument, wildlife refuge, marine protected area, flora micro-reserve and biosphere reserve are just very limited examples of the vast number of legal PA categories in use across the world. The IUCN proposed a reduced number of PA categories based on their management objectives with the aim of standardising the plethora of local, subnational, national and regional PA categories, facilitating mutual understanding and streamlining reporting to international organisations (Table 3.1).

PAs are not designated in isolation, but usually included in national or international PA systems or networks, depending on their degree of coordination. A good example of a highly internationally coordinated PA network using common designation, management, evaluation and reporting criteria is the Natura 2000 Network (EEC 1992). The Natura 2000 Network derives from the legislative work undertaken by the European Union shortly after the signature of the CBD, in 1992. Those efforts crystallised in the Habitats Directive, an international legally binding norm for the EU members establishing common criteria for the identification and adequate conservation of European biodiversity in terms of wild species and natural and semi-natural ecosystems (European Commission 2021a). The Habitats Directive created the Natura 2000 Network, which is made of three categories of PAs (Fig. 3.3): Sites of Community Importance (SCIs), Special Areas of Conservation (SACs) and Special Protection Areas for Birds (SPAs). Currently, there are 26,918 Natura 2000 sites covering 1,204,987 km^2 of the EU's 27 countries' land and sea areas (European Commission 2020a).

The important work developed by the European Union with the Natura 2000 Network has inspired some neighbouring non-EU countries in Eastern and northern Europe to adopt a similar conservation approach and criteria by creating the Emerald Network of Areas of Special Conservation Interest (Council of Europe 2021). Together, the Natura 2000 Network and the Emerald Network make the largest internationally coordinated *in situ* conservation effort of the world.

Sometimes, PAs are designated under just one legal category, for example, nature reserve. However, this is rarely the case, especially in old, highly developed PA networks or in large PAs harbouring a diversity of valuable conservation features, from species to landscapes. In these cases, it often happens that different legal categories of PAs partially or completely overlap spatially over a single space, likely providing greater legal protection through accumulation of restrictions to human

Table 3.1 Protected area management categories according to the International Union for Conservation of Nature

Category	Description
Ia. Strict nature reserve	Protected areas that are strictly set aside to protect biodiversity and also possibly geological/geomorphological features, where human visitation, use and impacts are strictly controlled and limited to ensure protection of the conservation values. Such protected areas can serve as indispensable reference areas for scientific research and monitoring
Ib. Wilderness area	Protected areas that are usually large unmodified or slightly modified areas, retaining their natural character and influence, without permanent or significant human habitation, which are protected and managed so as to preserve their natural condition
II. National park	Large natural or near natural areas set aside to protect large-scale ecological processes, along with the complement of species and ecosystems characteristic of the area, which also provide a foundation for environmentally and culturally compatible spiritual, scientific, educational, recreational and visitor opportunities
III. Natural monument/feature	Protected areas set aside to protect a specific natural monument, which can be a landform, sea mount, submarine cavern, geological feature such as a cave or even a living feature such as an ancient grove. They are generally quite small protected areas and often have high visitor value
IV. Habitat/species management area	Protected areas aiming to protect particular species or habitats and management reflects this priority. Many category IV protected areas will need regular, active interventions to address the requirements of particular species or to maintain habitats, but this is not a requirement of the category
V. Protected landscape/seascape	A protected area where the interaction of people and nature over time has produced an area of distinct character with significant ecological, biological, cultural and scenic value: and where safeguarding the integrity of this interaction is vital to protecting and sustaining the area and its associated nature conservation and other values
VI. Protected area with sustainable use of natural resources	Protected areas that conserve ecosystems and habitats, together with associated cultural values and traditional natural resource management systems. They are generally large, with most of the area in a natural condition, where a proportion is under sustainable natural resource management and where low-level non-industrial use of natural resources compatible with nature conservation is seen as one of the main aims of the area

uses in the area (Rodríguez-Rodríguez et al. 2016). For instance, in the European Union, it is common to have national parks that are also Natura 2000 sites, or large nature parks or biosphere reserves that include smaller natural monuments or nature reserves within their boundaries.

Even if a PA is designated under one legal category, it is also common that they are divided into management zones, especially if they are large, environmentally diverse or include human dwellings, infrastructures or other human interests inside them. In zoned PAs, there is typically an inner core zone that contains the most vulnerable biodiversity to protect and has strict restrictions to human uses, and a multiple-use zone surrounding the core zone. Multiple-use zones are usually large.

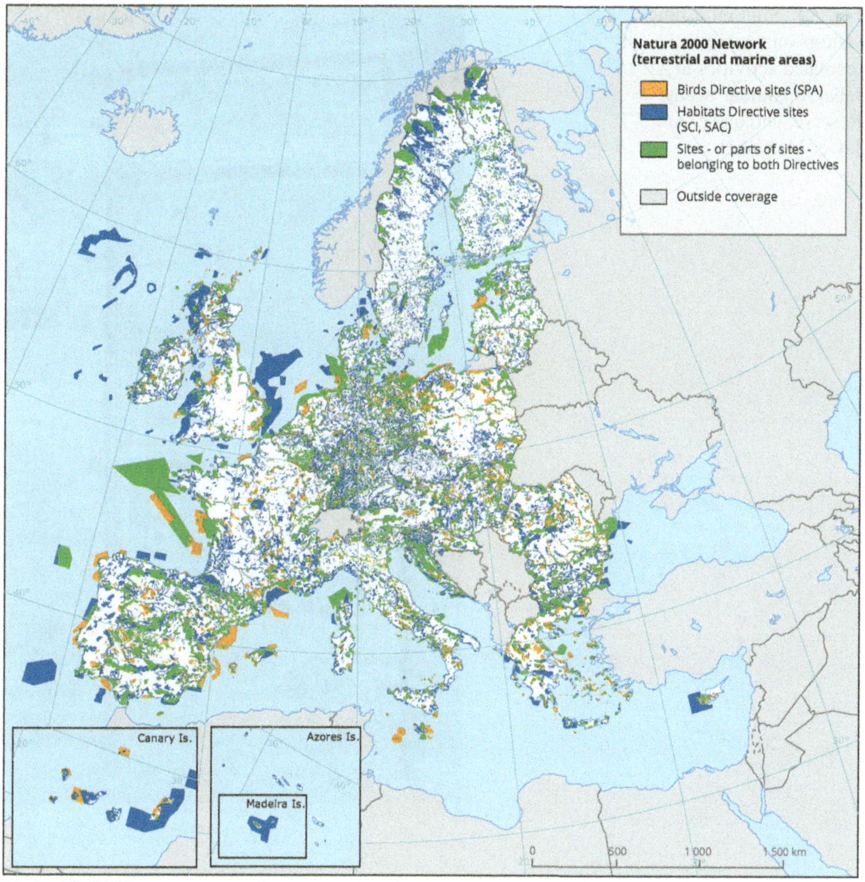

Fig. 3.3 Extent and typology of the Natura 2000 Network by September 2020. (*Source*: European Environment Agency https://www.eea.europa.eu/data-and-maps/figures/natura-2000-network-terrestrail-and)

They are intended to buffer external impacts to the most ecologically sensitive core zones while allowing a broader range of human uses that are considered compatible with the conservation of protected biodiversity such as farming, tourism or artisanal fishing.

Who makes decisions in PAs? The CBD identifies four governance types in PAs: governance by governments at different administrative levels; shared governance by diverse rights holders and stakeholders; governance by private entities; and governance by indigenous peoples and/or local communities (Borrini-Feyerabend et al. 2013). Decisions about PAs are normally implemented by PA managers, who can belong to any of the above-mentioned groups. PA management entails a range of regular activities typically including surveillance, ecological monitoring,

Fig. 3.4 Sign informing visitors on zoning and regulated activities in Al Jabal Al Akhdar Scenic Reserve, Oman. (*Credit*: Javier Martínez)

conservation actions, visitors' regulation and sites' administration. Most PA management is made by governments at different scales. Though management is explicit in both international PA definitions and largely desirable to achieve conservation objectives (Dudley 2008), many PAs around the world do not have management in place, which likely reduces their ability to deliver positive conservation outcomes (Xu et al. 2019). Management can be vital for the survival of threatened biodiversity in conflicting places by avoiding poaching or preventing land clearing and occupation. Many iconic PAs around the poor regions of the World would not maintain their valuable biodiversity if permanent dedicated staff did not make tremendous efforts to enforce conservation regulations, sometimes even risking their own lives (Global Conservation 2018).

However, insufficient funding prevents most unmanaged PAs from having regular or sufficient management resources (Wilkie et al. 2001). In developed countries most managerial funds come from governmental institutions (i.e. state funding), whereas developing countries depend on donations, entrance fees or direct involvement of charities and private entities to a much larger extent. For instance, most PAs in Madagascar are managed by national or international environmental NGOs and

Fig. 3.5 State-funded visitor centre of Mtirala National Park, Georgia. (*Credit*: David Rodríguez)

private organisations entrusted to protect the country's unique biodiversity on behalf of the country's Ministry of Environment and Sustainable Development due to the Ministry's insufficient capacity to do so. Those organisations apply a nationally regulated entrance fee plus compulsory guided tour scheme to PA funding with a dual local-foreigner pricing policy. Half the entrance fees support the Ministry's administration and coordination activities, whereas the other half is aimed at local development projects in the communities surrounding PAs. In turn, visitors to PAs are required to hire a local guide to tour them inside the country's PAs. That way, local communities benefit from PAs in terms of tourism revenues, and visitors' potential inappropriate behaviour (e.g. collecting wildlife; littering or vandalising geological features) is controlled and reduced. Well devised as such management and funding schemes may be, irresponsible behaviour, irregular use of funds and general scarcity of resources still hamper sustainable development in many PAs across the country (Méral et al. 2011).

PA management requires substantial resources that usually increase with size, remoteness, environmental complexity, conflicting and broad-ranging species, intensity and diversity of human uses. The larger and more remote a PA is, the more effort is needed to adequately survey, monitor and protect threatened biodiversity. Although remoteness may naturally protect some PAs from human pressures (Andam et al. 2008; Edgar et al. 2014), many large remote PAs desperately struggle to prevent illegal activities like land clearing, poaching and low-scale mining (Rija 2017; Wanghe et al. 2020). Limited managerial resources and difficulty to access some distant or inaccessible places make controlling such activities specially challenging in those areas (Chagos Conservation Trust 2021). Environmental diversity typically increases managerial costs too. Patrolling a homogeneous grassland PA is far simpler (and cheaper) than surveying a PA with a diversity of different ecosystems, including for instance wetlands, mountains, forests, coastal and marine areas, which most likely require different means and techniques (Lajeunesse 2021). PAs that protect conflicting biodiversity, especially large carnivores or other megafauna that may cause direct impacts to local human populations, such as elephants, also require substantial managerial resources as well as social skills to deal with clashing human-animal relationships. Such relationships are particularly relevant in developing countries where cattle losses due to predation or crop destruction are not normally offset by state's subsidies and may entail serious consequences for human well-being and survival (Karanth and Nepal 2012; Sekhar 1998). In those situations, managing wildlife impacts on human livelihoods so that retaliation in the form of trapping, poisoning or poaching of impacting species is avoided requires extraordinary talent and skills that may not always suffice. Finally, the larger the number of human activities developed in PAs or their surroundings and the greater their intensity, the more challenging PA management will be. For instance, PAs where indigenous populations have to cope with expanding cattle ranchers, mining companies and farmers are prone to severe conflict (Ferrante and Fearnside 2019). Human uses of PAs are tried to be made compatible through PA zoning, surveillance, enforcement and participatory governance. However, deficient social engagement and inefficient enforcement due to insufficient managerial means, unprofessional behaviour or corruption often reduce PAs' effectiveness, especially in developing countries (Schoneveld-de Lange et al. 2016).

PAs are complex territorial entities, and as such, they experience a range of threats and pressures that they are intended to reduce and that may limit their effectiveness. Two broad categories of pressures can be identified: natural pressures and human-made pressures. Natural pressures include regular or sporadic catastrophic events such as ray-produced wildfires, landslides, earthquakes, volcanic eruptions, hurricanes or floods. Biodiversity is usually adapted to regular such perturbations and normally recovers after some time. However, if such events are rare and/or have severe intensity (for instance a large volcanic eruption), they may become a very serious threat to local biodiversity (Mendoza et al. 2020).

Human-made pressures are, however, the most common and worrisome (Bowler et al. 2020; Halpern et al. 2008). Not only are they the most numerous and widespread, but also generally the most destructive ones. Pressures to PAs are highly context dependent, but there are some common pressures to PAs worldwide. Typical pressures on PAs include a range of socio-economic activities including farming, logging, infrastructure development, housing, hunting, fishing, harvesting, mining,

Fig. 3.6 Shared management schemes are common in developing countries. (*Credit*: David Rodríguez)

industrial activities and tourism that result in a large variety of impacts on PAs and wildlife (Mammides 2020). PAs are intended to protect sensitive biodiversity from those pressures, but inadequate regulations, deficient management or complex socio-economic contexts sometimes result in poor conservation outcomes. Moreover, even if those pressures are abated inside PAs, broad-ranging impacting activities occurring in their surroundings may severely affect biodiversity in PAs. An example is oil spills from oil rigs or oil tanker accidents near MPAs (Dalton and Jin 2010).

An often overlooked but yet highly relevant threat to PAs is political will. After all, protecting biodiversity is a human choice that usually entails a compromise between short-term economic development and long-term nature conservation that might be difficult to convey to local populations, especially in poor countries with pressing social needs or vested interests by the ruling elites. Such compromise has sometimes resulted in a dangerous outcome: the downgrading, downsizing or degazzetement of PAs (PADDD; Mascia and Pailler 2011). PADDD processes have

Fig. 3.7 Poaching is a major source of concern for many protected area managers, especially in developing countries. (*Credit*: David Rodríguez)

led to a loss of protected status of part or all of some PAs, or to a less stringent legal status, potentially affecting biodiversity in those sites seriously (Qin et al. 2019).

Thus, given the diversity of environmental, socio-economic, managerial and political issues that may affect PAs, can PAs still deliver effective conservation? Can they be effective in all contexts? What factors make them more effective at fulfilling their role of conserving biodiversity?

Chapter 4
Protected Area Effectiveness: What Is It and How to Assess It?

The Natura 2000 Network currently expands over 18% of the European Union's land territory (European Commission 2020a), thus exceeding global protection coverage targets for 2020 (CBD 2010). In contrast to this large conservation effort, it may shock some to know that 73% of the Union's species of conservation concern, including 57% of bird species, 82% of freshwater fish species, 72% of non-vascular plants and 75% of mammals in the Union, do not reach good ecological status. Even worse, just 15% of the natural and semi-natural ecosystems are in good ecological condition (European Commission 2020a). What is going on? How are these disappointing figures possible in a powerful, resourceful region with sound conservation legislation and a well-developed network of marine or terrestrial protected areas (M)PAs?

Different studies have shown that, despite the rise in PA numbers and coverage worldwide, the global deterioration of biodiversity has not been stopped (Butchart et al. 2010; WWF 2020). Are PAs adequately fulfilling their mission of conserving biodiversity? That is the key question that PA effectiveness assessment/evaluation seeks to answer. Before moving further into the issue of PA effectiveness, there are some key concepts to be known (Table 4.1).

PAs are not the panacea against the current biodiversity crisis. Generally, they have been found to slow the rate of deterioration of biodiversity when compared to absence of PAs, but complete stop or reversal of degradation is rare (Pfeifer et al. 2012; Geldmann et al. 2019). There are cases of PA conservation success and failure, although failures tend to be more rarely reported (Mora and Sale 2011).

Substantial financial, human and technical resources are spent every year to sustain PAs globally, so it is important to ascertain whether PAs are accomplishing their mission, not only on conservation grounds but also in terms of legislation use and economic efficiency (Fuller et al. 2010). Actually, as mentioned earlier, funding is a major hindrance to PA effectiveness. Most PAs are underfunded to carry out

D. Rodríguez-Rodríguez, J. Martínez-Vega, *Effectiveness of Protected Areas in Conserving Biodiversity*, Strategies for Sustainability,
https://doi.org/10.1007/978-3-030-94297-7_4

Table 4.1 Key concepts on protected area effectiveness

Concept	Definition	References
Assessment	The measurement or estimation of an aspect of interest	Hockings et al. (2006)
Evaluation	The judgment of the status/condition or performance of some aspect of interest against predetermined criteria (usually a set of standards or objectives); in this case including the objectives for which the protected areas were established	Hockings et al. (2006)
Monitoring	The regular and systematic collection of environmental and biological data by agreed methods and to agreed standards	Lockwood (2010)
Indicator	Biodiversity indicators are information tools that summarise data on complex and sometimes conflicting environmental issues to indicate the overall status and trends of biodiversity	Ten Brink (2006)
Protected area management	Active steps to conserve the natural (and possibly other) values for which the protected area was established; 'management' can include a decision to leave the area untouched if this is the best conservation strategy	Dudley (2008)
Management effectiveness	The assessment of how well the protected area is being managed – primarily the extent to which it is protecting values and achieving goals and objectives	Hockings et al. (2006)
Protected area effectiveness	Capacity of protected areas to conserve their resources in the long term	Rodríguez-Rodríguez (2012)

Literal or adapted from the original source

essential management activities, especially in developing countries (Mora and Sale 2011; Cook et al. 2019). In the early 2000s, Bruner et al. (2004) estimated that existing PAs in developing countries were running short of funding by 1 billion USD to 1.7 billion USD per year and that adequately managing an expanded, representative PA system in those countries would require up to 4 billion USD annually until 2020. New, ambitious post-2020 targets (CBD 2018) will most likely increase the number of PAs and the area under protection schemes substantially, further straining available management resources. Balmford et al. (2004) calculated that a global MPA network covering 20–30% of the global ocean would cost between 5 and 19 billion USD yearly to run while creating around 1 million jobs. Although the economic benefits of conserving biodiversity are estimated to substantially exceed the entailed costs (Naidoo and Adamowicz 2005; Yao et al. 2019), it is necessary to optimise both natural and economic resources in a context of foreseeable rising scarcity. In the debate over PA quantity versus quality, some authors state that limited expenditure on PAs should prioritise management of existing PAs to designation of new ones on cost-effectiveness grounds (Adams et al. 2019). Others affirm that replacing ineffective or poorly located PAs with new PAs that include the most representative places to biodiversity can entail a massive step for conservation without incurring extra costs (Fuller et al. 2010).

The IUCN developed a widely influential PAME Framework that conceptualised PA management effectiveness (PAME) for the first time (Hockings et al. 2000, 2006). It divided PAME evaluation into six elements defining the management cycle: context, planning, inputs, processes, outputs and outcomes (Fig. 4.1).

In their extensive review, Leverington et al. (2010) found over 70 PAME methodologies in use around the world, many of them influenced by the PAME Framework (Hockings et al. 2006). PA evaluation methods can be classified according to the type of data used as:

1. *Qualitative, opinion-based methods*. Such methods typically use primary data from surveys or interviews to PA managers or other stakeholders. They allow for rapid, cost-effective assessments although their accuracy and precision are improvable (Carranza et al. 2014a; Cook et al. 2014). They are the largely prevailing PA assessment/evaluation methods, especially in developing countries (Leverington et al. 2010). The two most widely used PAME methods belong to this category: RAPPAM (Ervin 2003) and METT (Stolton et al. 2007).

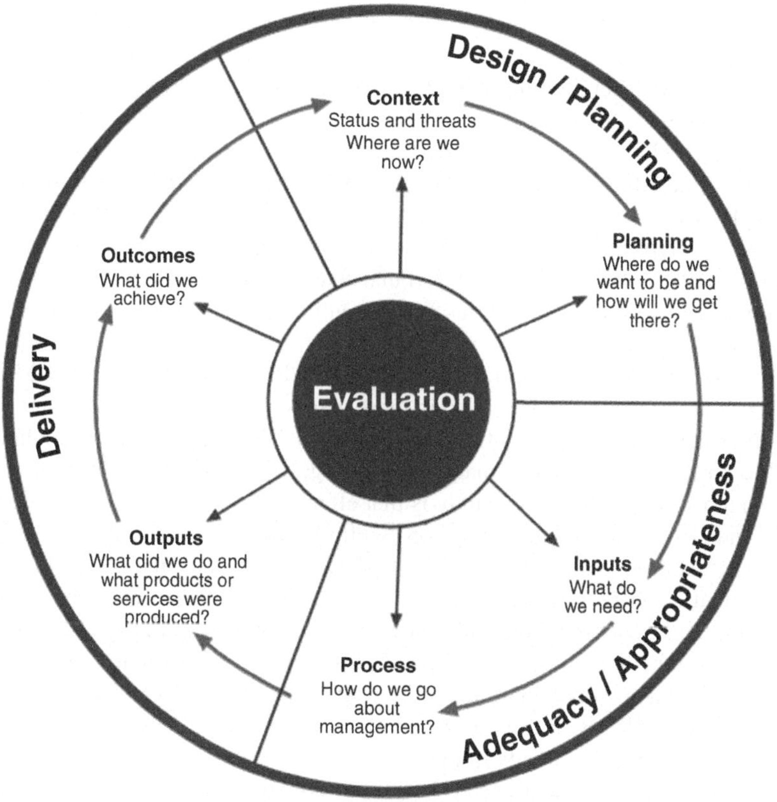

Fig. 4.1 Protected area management cycle proposed by Hockings et al. (2006)

2. *Quantitative methods*. They use statistical data from primary or secondary sources including on-site biodiversity monitoring, official statistics and remote sensing. They usually provide high accuracy and precision, especially if carried out according to standardised data collection protocols (Addison 2011). However, they are rarely used in (M)PA effectiveness assessments, even in data-rich countries (Addison et al. 2015). An example of a well-designed biodiversity quantitative monitoring system is that in force in the European Union's Natura 2000 Network. It aims at providing accurate estimates on the status and trends of species and habitats of conservation concern at least every 6 years (European Economic Community 1992). However, insufficient data completeness and inadequate on-site evaluation methods where expert judgment prevails (European Commission 2021b) limit the full potential of such a system.
3. *Mixed methods.* They combine quantitative, objective data with subjective, qualitative data from surveys or interviews. The SIAPA (Rodríguez-Rodríguez and Martínez-Vega 2012) is one of such methods.

All these methods typically use a range of effectiveness indicators (Gaston et al. 2008) that can be just measured (in assessments) or measured against reference target values (in evaluations) to enhance interpretation of results. They can be used to assess/evaluate only PAs or use more complex research designs that compare results from the assessment/evaluation in PAs with one or several control sites chosen according to different criteria to increase the validity of assessments (Addison 2011; Andam et al. 2008).

The work by Leverington et al. (2010) was used to create the Global Database on Protected Area Management Effectiveness (GD-PAME), which seeks to record existing PAME initiatives and their results around the world. GD-PAME is linked to the World Database on Protected Areas (WDPA) and managed by the UNEP-WCMC (2010). In their review, they found that approximately 9000 site assessments had been done in 140 countries by 2010, although that quantity only amounted to 6% of all registered PAs in the WDPA at the time (Leverington et al. 2010). Rapid, opinion-based methodologies such as RAPPAM (Ervin 2003) and METT (Stolton et al. 2007) widely prevailed in those assessments. After standardising data from different sources, their meta-analysis of 4151 PAME assessments resulted in just 24% of them having sound management, largely in developed countries.

An interesting result of their study is that effective PA management was significantly and strongly correlated ($r > 0.65$) with 'Adequacy of infrastructure, equipment and facilities' and 'Effectiveness of administration, including financial management', whereas it was weakly correlated ($r < 0.39$) with the current 'Condition and trend of nominated values' (Leverington et al. 2010). In contrast, conservation outcomes were chiefly correlated with the 'Constraint or support by external political and civil environment' ($r = 0.42$), a 'context' variable external to the PA and, secondly, with 'Research and monitoring' ($r = 0.35$), two purely managerial activities linked to inputs and processes.

The most frequently mentioned criterion for effective conservation in PAs is PA management (Gaston et al. 2008), which even gave its name to the current PA management effectiveness (PAME) research field (Hockings et al. 2000). But does effective management equal effective conservation?

A well-managed PA is more likely to deliver better conservation outcomes than a poorly managed PA or an unmanaged one (Kearney et al. 2020). However, PA effectiveness is multifactorial and highly context dependent. Thus, whereas the former statement is generally true (Barnes et al. 2016; Edgar et al. 2014), effective management may be neither sufficient nor even needed for effective conservation in cases where strong external pressures or, inversely, absence of pressures (e.g. inaccessible PAs in remote places) makes management a rather limited input to conservation (Gaston et al. 2008). A number of studies around the world have evidenced that conservation outcomes does not seem to be that much linked to PA management (Nolte and Agrawal 2013; Rodríguez-Rodríguez et al. 2015a), but to a mix of 'external' variables (i.e. context) and 'internal' ones among which enforcement and management resources rank high, especially in developing countries and in the marine environment (Edgar et al. 2014; Schleicher et al. 2019). Key threats to PAs, such as LULC changes or PADDD processes, are largely regulatory, governance and political decisions that have been seldom considered to this date (Tesfaw et al. 2018; Watson et al. 2014). Therefore, having enough managerial resources for monitoring and enforcement is usually necessary though not sufficient to provide effective conservation in all PAs.

As a result, though the PAME Framework (Hockings et al. 2006) was clearly a step ahead at the time and useful at conceptualising management effectiveness, it may nominally and practically be misleading PA effectiveness assessments by missing the focus on what really matters: the actual biodiversity outcomes of PAs. Current PAME does so by combining 'external' context variables that are largely outside the scope or remit of PA managers such as some threats to PAs, legal or political issues (Gaston et al. 2008), with purely managerial variables to which its name refers. Thus, it has been proposed to differentiate PAME, restricted to the evaluation of purely managerial activities, from PA effectiveness, which focuses exclusively on conservation outcomes, which may result from a diversity of external and internal factors to PAs (Rodríguez-Rodríguez 2012).

Building on the PAME Framework and the growing body of evidence differentiating PA management effectiveness and PA conservation effectiveness, the IUCN recently developed and put forward the Green List, an updated standard methodology and certification scheme for effective PA conservation that consistently assesses PA performance using 50 indicators under 4 main components: governance, design and planning, effective management and conservation outcomes (IUCN 2017). Currently, 59 sites in 16 countries have been awarded the Green List label (IUCN 2021a) (Fig. 4.2).

Current data from the Global Database on PAME shows that although some (M) PA effectiveness evaluation has been done in 169 countries, only 11% of the world's (M)PAs covering 18.3% of the global protected area had been evaluated by 2020, chiefly using subjective self-assessment methodologies (Protected Planet 2020). Moreover, the globally agreed indicator on PAME under Aichi Target 11 (CBD 2010) just assesses whether an (M)PA has undergone any assessment on its

Fig. 4.2 Conceptual outline of the Green List assessment and certification scheme. (*Source*: IUCN 2017)

effectiveness, not the actual outcomes of that assessment, which are not always publicly reported (Hockings et al. 2006). If biodiversity conservation is to be streamlined, global (M)PA effectiveness must be improved. For this to happen, there is an urgent need for widespread, sound, consistent and regular implementation of (M)PA effectiveness evaluations of environmental outcomes based on reliable and updated monitoring data. Scant, opportunistic monitoring and evaluation can detect some spatial and time-specific gaps and trends, but could hardly be the basis for long-term efficient conservation actions.

Chapter 5
A Review of Protected Area Effectiveness Around the World: Methodological Approach

We used a systematic literature review (SLR) as a rigorous, standardised methodology to compile existing scientific evidence on the conservation effects of PAs in a comprehensive, consistent and reproducible manner (Okoli and Schabram 2010). We developed a written review protocol aimed at enhancing methodological consistency at selecting relevant articles for review and collecting data from the reviewed articles. Our protocol followed the eight steps set up as best practice in systematic literature reviews (Okoli and Schabram 2010).

5.1 Purpose of the Literature Review

We sought to respond to five thematic research questions:

1. Are PAs effective at conserving biodiversity?
2. What indicators, realms, biodiversity types, regions and countries have been assessed most?
3. What are the most relevant pressures to (M)PAs?
4. What factors contribute to the success or failure of PAs at conserving biodiversity?
5. Are there any positive and negative case studies to draw lessons from?

    An additional methodological research question was included:

6. What are the methods most widely used to assess PA effectiveness?

D. Rodríguez-Rodríguez, J. Martínez-Vega, *Effectiveness of Protected Areas in Conserving Biodiversity*, Strategies for Sustainability,
https://doi.org/10.1007/978-3-030-94297-7_5

5.2 Protocol and Training

In this phase, we explicitly established and agreed on the methodological requirements for our review, including the following points.

5.2.1 Search Database

We used Scopus as our search database given that WoS did not provide updated data for some of its thematic databases that were relevant to our search such as Agriculture, Biology & Environmental Sciences (ABES), Physical, Chemical & Earth Sciences (PCES) or Life Sciences (LS).

5.2.2 Data Sources

Peer-reviewed scientific articles in any Scopus-indexed journal were our target data sources. We excluded other data sources such as reports or grey literature on difficulty for comprehensive retrieval and information quality grounds. By exclusively using peer-reviewed scientific articles, we could ensure adequate quality standards of the retrieved (and eventually analysed) information.

5.2.3 Language

Only articles written in English were included in the review.

5.2.4 Date Range

We searched for all available articles that had been published in a decade, from 2010 until 2019.

5.2.5 Keywords

We searched for a number of relevant keywords in the articles' titles. We tested two search strings to ensure a balance between comprehensiveness and manageability of information:

- String 1 (broader): ('protected area' or 'MPA' or 'reserve' or 'natura 2000 site' or 'park') AND ('effect' or 'impact' or 'effectiveness' or 'performance' or 'efficacy')
- String 2 (narrower): ('protected area' or 'MPA' or 'reserve' or 'natura 2000 site' or 'park') AND ('effect' or 'impact' or 'effectiveness' or 'performance' or 'efficacy') AND ('biodiversity' or 'gene' or 'genetic' or 'species' or 'ecosystem' or 'habitat' or 'land' or 'environment*')

5.2.6 *Data to Be Retrieved*

An Excel data collection form containing the articles' main identification features as well as fields that responded to the research questions was produced (Table 5.1). The form was updated 11 times during the whole review process for comprehensiveness and clarification.

We assessed methodological consistency by testing the degree of inter-reviewer agreement on article selection on 3 different sets of 25 articles. We had an initial training session using the original version of the protocol, and 3 additional joint training sessions following the mutual reviews of the 3 sets of 25 identified articles from the original search and subsequent modifications of the protocol. Trainings were aimed at improving consistency of review decisions and complementing and clarifying the review protocol. Our degree of agreement on the selection of articles for review in the three rounds of review ranged from 76% ($\kappa = 0.35$) to 80% ($\kappa = 0.42$), showing a fair to moderate degree of agreement. Previous studies on SLR found poorer degrees of agreement in article selection processes between two reviewers after similar rounds of training (Staples and Niazi 2007).

5.3 Literature Search

Search string 1 produced 2787 scientific articles. We tried to narrow the article list down by including new discriminating terms in the search. String 2 produced a more targeted, manageable sum of 411 articles which were screened for inclusion in the review.

5.4 Article Screening

We split the final article list ($N = 411$) so that each author reviewed half of it. Two consecutive screenings were performed on each sublist. Firstly, a quick screening based on the articles' titles was used to initially refine each sublist. Secondly, a more detailed screening based on the articles' abstracts provided the final set of articles

Table 5.1 Data collection form

Data type	Field name	Description	Field values
Article ID	Publication code	Article-specific, ordinal code of selected articles for retrieval purposes	Consecutive ordinal numbers
	First author	Full name of the first author	Name(s) & surname(s)
	Journal	Journal where the article was published	Journal name
	Publication year	Year where the article was published	2010 till 2019
	DOI	Article's DOI	DOI code
Review data	Biodiversity	Type of biodiversity addressed by the article	Genes; species; ecosystems[a]
	Taxa	Classification of the studied genes/species	Fungi; plant; invertebrate; fish; amphibian; reptile; bird; mammal
	Realm	Major environment where the study took place	Terrestrial; freshwater; marine; mixed (coastal)
	Indicators used	Variable(s) used to assess/evaluate PAs	For example, abundance of carnivore species
	Indicator size	Number of biodiversity units being measured	For example, ten non-flying bird species; eight LULC classes
	PA category	Legal category of the studied PAs	For example, biosphere reserve; nature park If the site has >1 legal category (e.g. nature park and biosphere reserve), only the category portrayed by the authors will be considered
	PA type	Generic legal stringency of the PA category assessed: high stringency (Reserve) or moderate stringency (multiple-use)	Reserve (e.g. national park; nature reserve); multiple-use (e.g. Natura 2000; nature park)
	Number of PAs	Number of PAs being assessed or evaluated	N (whole PA network); n (sample of PAs)
	Research design	Generic classification of the research design used for the assessment based on the 'hierarchy of evidence' (Burns et al. 2011)	1. Experimental (random, case-control[b] treatment) 2. Complete semi-experimental (BACI) 3. Incomplete semi-experimental (no controls and/or no before data) 4. Mixed (different designs for different indicators)
	Assessment/ evaluation method	Specific assessment method from research design	M-P-B-A-C-I
	Assessment technique	Methodological tools used to gather assessment data	For example, remote sensing (RS); in situ sampling; literature (maps; registries; books, etc.)

	Time range	Number of years in which indicator data were collected	1. Broad (time series[c] spanning >10 years, even if non-consecutive) 2. Moderate (time series ≤10 years) 3. One-off (1 point in time) Time range will be specified
	Threshold	Existence of compliance threshold(s) for the indicators used	Yes: Evaluation; no: assessment; mixed
	Region	World's region (continent) where the study took place	Africa; America; Antarctica; Asia; Europe; Oceania; Global
	Country	Country(ies) where the study took place	Name of the country
	Development	Study country(ies)' human development ranking	Very high; high; medium; low[d]
	Outcome	Result of the assessment/evaluation	*1. Positive* The outcome will be *positive* if all the PA sample and/or indicators show better values with or after the PA *2. Slightly positive* It will be *slightly positive* if most of the PA sample and/or indicators show better values with or after the PA *3. Negative* The outcome will be *negative* if (1) all the PA sample and/or indicators show worse values with or after the PA or (2) the indicators show no difference in value with regard to control areas[e] *4. Slightly negative* The outcome will be *slightly negative* if most of the PA sample and/or indicators show worse values with or after the PA *5. Mixed* The outcome will be *mixed* if roughly half of the sample of PAs and/or indicators show positive and negative values with or after PAs *6. Inconclusive* The outcome will be *inconclusive* if the authors state so or there is not enough evidence to judge If outcomes can be differentiated between PA types, they will be

(continued)

Table 5.1 (continued)

Data type	Field name	Description	Field values
	Indicator trend	Trend in the value(s) of the indicator(s) used	1. *Improving*: mostly better values at the end than at the beginning of the time range of the study 2. *Stable*: similar values at the end of the study 3. *Deteriorating*: mostly worse values at the end of the study 4. *Mixed*: some joint trends above 5. *Non-applicable* (one-off studies) If indicator trends can be differentiated between PA types, they will be
	Main pressures	Main pressures to biodiversity or PAs mentioned	For example, human development; poaching
	Effectiveness factors	Identified factors that contribute to PAs' effectiveness or lack of effectiveness	Factors contributing to PAs' effectiveness (+) or lack of effectiveness (-) that are mentioned by the authors
	Recommendations	Authors' recommendations to improve PA effectiveness	For example, investing more in surveillance; more research
	Case studies	Relevant examples of PA effectiveness success and failure	Yes (showing clearly positive or negative outcomes); no

[a] *Biodiversity* field remark: Land uses-land covers (LULC) were used as proxies for ecosystems

[b] *Research design* field remarks: With controls only outside protected areas; If the first indicator data came from the same year when the PA was designated, they were considered 'Before' designation baseline data, and the method will be considered BACI

[c] *Time range* field remark: Two data from different dates were considered the minimal time series

[d] *Development* field remark: According to UNDP 2020. *Human Development Report 2020. The Next Frontier: Human Development and the Anthropocene*. New York. United Nations Development Programme. hdr.undp.org/sites/default/files/hdr2020.pdf (Retrieved 25 January 2021)

[e] *Outcome* field remark: Null and truly negative effects were thus considered equivalent to better reflect lack of effectiveness of (M)PAs, as done in previous studies (Geldmann et al. 2013)

for review. In both screenings, if in doubt, articles were moved to the next phases of the review (Okoli and Schabram 2010).

5.5 Quality Appraisal

Okoli and Schabram (2010) recommend appraising pre-selected documents for quality before retrieving data from them. Whereas that is a very sensible recommendation in broad SLRs that include all sources of information including grey literature, we skipped the 'quality appraisal phase' as we assumed that all the articles in journals in Scopus had been peer-reviewed before publication which ensured an acceptable standard of quality. Thus, it was outside our scope to act as additional reviewers of the pre-selected articles.

5.6 Data Extraction

The set of pre-selected articles from the second screening phase was split in two subsets and scanned by each reviewer to supply data to the data form. The articles' Methods, Results, Discussion and Conclusion sections were carefully looked at, as well as the final parts of the Introduction (objectives of the study), for being the specific, data-richest parts of the studies. Supplementary data were not reviewed.

Some inclusion and exclusion criteria of articles were progressively agreed at this phase:

1. Articles on impacts of different factors on PAs' biodiversity were assessed in terms of effectiveness (or lack of effectiveness) if they might refer to legal or managerial limitations. We assumed any such improvement or deterioration of biodiversity in PAs in terms of effectiveness.
2. Articles on impacts of different factors on PAs' features other than biodiversity (e.g. water quality) were not considered.
3. Studies that only used future or past predictions (scenarios; modelling; EIA) were not considered.
4. Studies comprising only gap analyses (i.e. representation of biodiversity in a PA system) were excluded, as they do not ascertain conservation effectiveness.
5. Purely ecological studies (e.g. habitat use; species diversity across habitats, etc.) within PAs that do not include population trends (i.e. using PA just as the context of the study with no controls in terms of sites' protection) were also excluded.
6. Studies focusing only on PADDD processes were also excluded.
7. Studies that were unclear on methods and/or results were also excluded in order to minimise interpretation by the reviewers.

8. One-off studies carried out only inside PAs were discarded for not having Before data, Control data or, at least, a time series of inside (Case) data (i.e. After-Impact) for comparison and assessment.

When 'Outcomes' for the same indicators were assessed using different methods, we only annotated the most valid one according to the hierarchy of evidence (Okoli and Schabram 2010; e.g. remote sensing data before interview data).

Once work by both reviewers concluded, we merged both sub-list assessments into a single Excel table for analysis and synthesis. Appendix 1 includes all the reviewed articles.

5.7 Synthesis

Results from the selected set of articles ($n = 76$) were summarised according to the research questions and fields in the data form. Descriptive statistics were computed for each field. The data compiled for open fields such as 'Indicators used', 'Effectiveness factors' or 'Pressures' were aggregated in common, related categories or headline indicators, and descriptive statistics were also computed for them, as done previously (Leverington et al. 2010). For instance, for the variable 'Indicators used', the indicators 'habitat loss', 'land development', 'deforestation', 'urbanisation' and 'LULC change' were all included under the common headline indicator 'LULC change'.

Some fields in the data form were coded and statistically analysed for differences and correlations in 'Outcomes' and 'Indicator trends' using non-parametric tests after testing for normality. Kruskal-Wallis tests and Spearman rank-order correlation tests, both for $\alpha = 0.05$ were used. Inferential statistics would not be needed using a census sample of articles that met our selection criteria, as we did. However, we used such tests to delve into possible linkages and relationships among variables for deeper explanation of results, in this case theoretically assuming that our reviewed articles came from a random sample of articles that met our selection criteria from a wider universe of articles on (M)PA effectiveness.

Finally, eight case studies were selected from the reviewed articles taking into account the following selection criteria:

1. Biodiversity *outcomes* (cases showing clearly positive or negative outcomes)
2. *Realms* (a balanced representation of realms: marine, coastal, terrestrial and freshwater sites)
3. *Biodiversity* (an adequate representation of species and ecosystems)
4. *Type of PAs* (2 marine reserves, 1 nature reserve, 3 national parks, 1 regional park and 1 mixed case incorporating all protection categories, from IUCN categories I–VI)
5. *Regions* (2 America, 2 Europe, 2 Africa and 2 Asia)
6. *Degree of development of countries* (2 very high, 4 high, 1 medium and 1 low; UNDP 2020)

5.8 Writing the Review

One of the reviewers wrote the bulk of the review to ensure consistency in style, whereas the other reviewed what was written to maximise accuracy, as suggested (Okoli and Schabram 2010).

5.9 Methodological Considerations

Research results from SLRs are not without limitations. SLRs are considered more structured, comprehensive and reliable than traditional, unstructured literature reviews (Massaro et al. 2016), although criticism to those assumptions has been made (Boell and Cecez-Kecmanovic 2015). An important source of potential quality concern of the results from SLRs is the error-prone nature of SLR itself. The vast amount of information to be reviewed, the different methods and terms used by the articles' authors, the long time needed to undertake a SLR (van Dinter et al. 2021) and the numerous variables, criteria and thresholds agreed upon by the reviewers (or even to be considered by just one reviewer) make completing an error-free SLR a very challenging task.

From our experience, we can conclude that SLRs, while being more structured, comprehensive and reliable than traditional reviews, cannot (and should not) claim absolute comprehensiveness or reliability. In spite of a number of factors facilitating review's coherence, including a regularly updated review protocol, three training sessions, the fact that we were just two reviewers and that we had worked on the same topic with each other for many years, we have found that adhering to equal selection and evaluation criteria was challenging and could not be guaranteed. Individual differences in backgrounds, expertise, interests and beliefs are likely to influence article selection and data acquisition from a moderate to large extent. As previously observed, studies that are complex, that portray mixed results that are difficult to interpret or that lie far from the reviewers' field of expertise have a greater chance of being left out of the review or inconsistently assessed (Boell and Cecec-Kecmanovic 2015). It is most likely that consistency worsens as the number of reviewers engaged in SLRs increases.

Some additional factors that might have contributed to different judgments on the reviewed materials include our prior limited experience with SLRs, the relatively long time to undertake the review, insufficient attention to the protocol and its 11 updates or the complexity of the topic itself. Assessing the effectiveness of PAs (whether PAME or purely their ecological effectiveness) is a relatively new field of research (Hockings et al. 2000). Therefore, a well-established consensus on methods, indicators and even terms to carry out (M)PA assessments has not yet been reached, although some international efforts are being done (Hockings et al. 2006; IUCN 2019). As a result, accurately assessing (M)PA effectiveness is a challenging task, with numerous indicators and approaches used, thus reducing internal consistency and comparability across sites (Leverington et al. 2010; Protected Planet 2021). Moreover, lack of ecological or managerial effectiveness thresholds greatly reduced analyses of (M)PA effectiveness in our SLR to assessments instead of more policy and managerially meaningful evaluations (Hockings et al. 2006).

The quality of results from SLRs is also highly determined by the quality of the data from the reviewed studies (Okoli and Schabram 2010). We reduced this caveat by resorting exclusively to peer-reviewed articles. However, the quality of the raw data compiled by the authors or of their results was assumed but could not be ensured. Less than one third of the reviewed studies applied reliable complete semi-experimental designs (BACI designs; Addison 2011), and just some of those used highly accurate 'matched' controls instead of default surrounding buffer areas. Almost one fifth of all the studies used little reliable *After-Impact* studies, and over 20% were one-off, in-out studies showing a snapshot of the conservation situation. This resulted in improvable accuracy and a likely overestimation of (M)PA effectiveness from the complete dataset (Andam et al. 2008). These results suggest the need to streamline scarce (M)PA assessment resources by using only highly valid BACI designs with biophysically similar controls and regularly replicate assessments in the same (M)PAs using the same research designs and indicators, so that biodiversity trends can be soundly produced, correctly interpreted and effectively reacted upon.

Chapter 6
Main Findings on Protected Area Effectiveness

6.1 Main Findings on Protected Area Effectiveness

A detailed review of 76 articles assessing 1,221,798 (M)PAs worldwide was made. In this section, we will summarise the results from our SLR that respond to the following thematic research questions:

6.1.1 Are PAs Effective at Conserving Biodiversity?

PAs were mostly effective at conserving biodiversity. 51.32% of the reviewed studies found that (M)PAs had had *positive* or *slightly positive* biodiversity outcomes. In contrast, for 26.32% of the studies, (M)PAs had had *negative*[1] or *slightly negative* biodiversity outcomes, whereas 19.74% of studies resulted in *mixed* outcomes, and 2.63% of studies produced *inconclusive* outcomes (Fig. 6.1).

When only highly reliable studies using complete semi-experimental research designs were considered ($n = 22$), (M)PAs were still mostly effective, although the difference in effectiveness between studies showing *positive-slightly positive* outcomes (45.45%) and those portraying *negative-slightly negative* ones (36.36%) was notably reduced (Fig. 6.1). BACI-type studies are thought to be the most reliable research designs producing the most internally valid results (Addison 2011). This result aligns with those from previous studies that reported reduced PA effectiveness in decreasing impactful LULC changes when using biophysically similar

[1]According to the protocol and previous studies (Geldmann et al. 2013), a study outcome was considered negative if the (M)PA could show no effectiveness either by (1) all the PA sample and/or indicators showing worse values with or after the PA or (2) the indicators showing no difference in value with regard to control areas.

D. Rodríguez-Rodríguez, J. Martínez-Vega, *Effectiveness of Protected Areas in Conserving Biodiversity*, Strategies for Sustainability,
https://doi.org/10.1007/978-3-030-94297-7_6

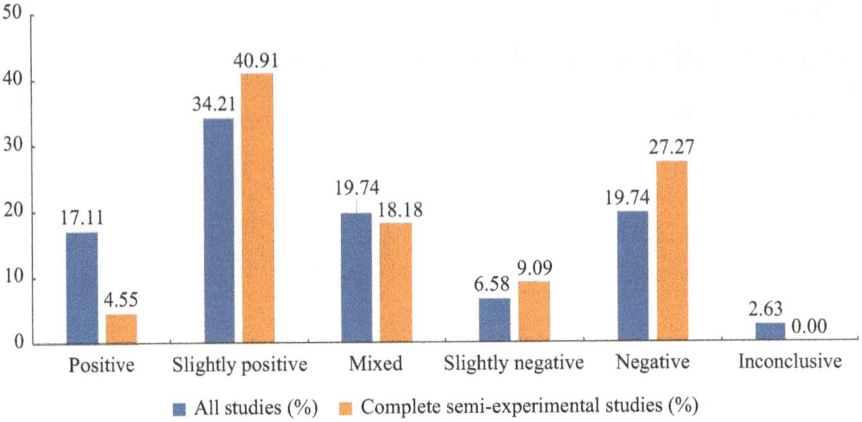

Fig. 6.1 Biodiversity outcomes of (M)PAs from the whole review (*n* = 76; blue columns) and from studies using complete semi-experimental research designs (*n* = 22; orange columns)

counterfactuals, the reason being that (M)PAs tend to be located in areas with lower pressures than the rest of the territory and are thus 'naturally' protected, for instance, by being located at high altitudes or on steep slopes (Andam et al. 2008; Joppa and Pfaff 2010).

A number of previous studies have likewise found generally positive outcomes of (M)PAs, although not in every case, or for every effectiveness indicator used (Geldmann et al. 2013; Coetzee et al. 2014, 2020). A worldwide review on PAME on over 3000 (M)PAs mostly based on qualitative data from interviews to (M)PA managers found similar moderately positive perceived results on conservation outcomes, scoring 0.56 in a zero to one scale (Leverington et al. 2010a, b).

Numerous studies found that (M)PAs were moderately effective in reducing habitat degradation and loss but did not completely avoid it (Andam et al. 2008; Spracklen et al. 2015). Other ecosystem-level studies showed less positive outcomes, especially for multiple-use PAs (Pfeifer et al. 2012) and non-forested PAs (Geldmann et al. 2020), whereas few studies reported negative conservation outcomes on ecosystems (Liu et al. 2001; Wang et al. 2018), although bias towards chiefly reporting significant and positive outcomes from (M)PAs has been reported and might be expected (Mora and Sale 2011) (Fig. 6.2).

Similarly to ecosystems, a number of studies showed generally positive conservation outcomes on species, even if on occasions mediated by avoided habitat degradation (Gillingham et al. 2015; White et al. 2017; Cazalis et al. 2020). In their literature review, Coetzee et al. (2014) found that species' abundance, assemblage abundance and richness were higher in PAs for invertebrates (arthropods), reptiles, birds and mammals, but similar to controls for plants. They also found greater abundance of threatened species inside PAs, except for critically endangered species which might result from their low sample sizes (Coetzee et al. 2014). Soykan and Lewison (2015) also consistently found greater species' abundance and biomass in MPAs, but no difference in species' richness and diversity with regard to control

Fig. 6.2 Protected areas reduce habitat degradation. Plitvice National Park, Croatia. (*Credit*: Javier Martínez)

areas. However, other studies found (M)PAs' effects on large carnivores to be uncertain and variable (Terraube et al. 2020), with another study even finding steady declines in large mammal populations in 78 African PAs between 1970 and 2005 (Craigie et al. 2010). Given that these authors used no controls, actual (relative) PA effectiveness could not be ascertained. Nevertheless, their results on absolute PA effectiveness are worrisome.

The effectiveness of (M)PAs is confirmed for a number of relevant variables in our SLR: *type of (M)PAs*, *realm*, *biodiversity* and *region*. All the categories had more positive than negative biodiversity outcomes, except for 'Asia' (*region*). For instance, 53% of all the assessed *reserves* had positive or slightly positive biodiversity outcomes, compared to 32% which had negative or slightly negative outcomes (Fig. 6.3).

Contrary to what was expected due to their greater legal stringency (Dudley 2008) and often broader managerial resources, reserves had just slightly better biodiversity outcomes than multiple-use (M)PAs. Previous studies found greater effectiveness of reserves over multiple use (M)PAs (Edgar et al. 2014; Rodríguez-Rodríguez and Martínez-Vega 2018a, b), whereas some others stated no difference in PA effectiveness across legal stringency categories (Nagendra 2008) or even greater effectiveness of multiple-use PAs (Coetzee et al. 2014).

By realm, studies showing positive outcomes approximately doubled those that had negative outcomes for every realm. (M)PAs in all realms were reported to have similar effectiveness proportions, the greatest being for MPAs and the smallest for

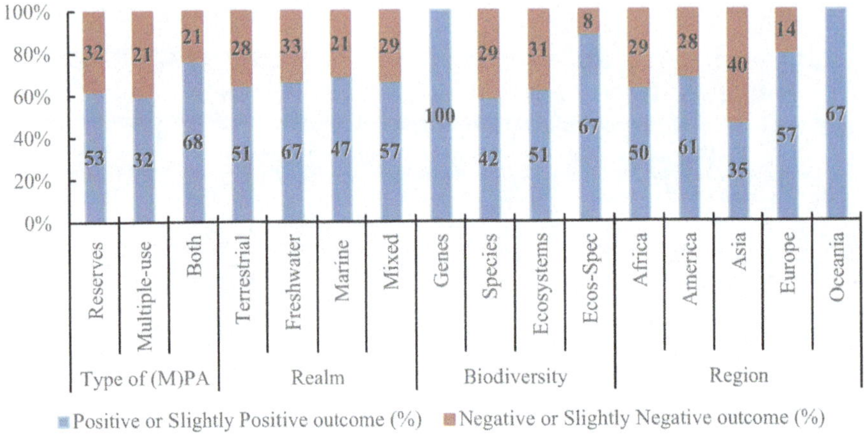

Fig. 6.3 Proportions of positive-slightly positive outcomes versus negative-slightly negative outcomes (in percentage) referred to 100% by category for four variables: type of (M)PA, realm, biodiversity and region

terrestrial PAs. The greater degree of spatial overlap of pressures in terrestrial environments (Bowler et al. 2020) is likely to reduce PA effectiveness in this realm (Fig. 6.4).

All biodiversity types showed largely more positive than negative outcomes. Apart from genetic biodiversity, represented in the review by just one study, the largest difference in outcome was from studies assessing both: species and ecosystems. Of all the studies showing only *positive* ($n = 7$) or *negative* outcomes ($n = 6$) for species or genes, invertebrates and mammals had more positive than negative outcomes from (M)PAs, whereas plant and fish had more negative than positive outcomes (Table 6.1). These results are similar to those by Coetzee et al. (2014) which in turn did not include fish and showed no conservation effects of PAs on plants.

Finally, nearly all regions had more positive than negative biodiversity outcomes. The best performing regions were Oceania (100% positive outcomes; $n = 3$) and Europe ($n = 21$), whose positive outcomes largely outnumbered negative ones. On the contrary, the studies on Asian (M)PAs reporting negative or slightly negative outcomes exceeded those with positive or slightly positive outcomes (40–35%, respectively). These results largely align with concerning outcomes on deforestation in Asian PAs by Nagendra (2008) and Spracklen et al. (2015) as well as with poor PA outcomes on avoiding combined pressures in the region (Geldmann et al. 2020). In contrast, Coetzee et al. (2014) found that PA effectiveness on species' conservation was the highest in Asia and Europe.

Some obvious conclusions arise from our findings. Firstly, (M)PAs are a useful policy instrument for conserving biodiversity globally. Thus, international agreements and national conservation policies should continue to have (M)PA expansion and effective management as the main goals for the years to come (CBD 2020).

Fig. 6.4 Protected areas at sea achieved the most positive biodiversity outcomes overall. (*Credit*: Pixabay (CC). https://pixabay.com/es/photos/peces-enjambre-submarino-1656504/)

Table 6.1 Sign and proportion of reported outcomes of PAs on species or genes from studies having positive ($n = 7$) or negative ($n = 6$) PA effectiveness results, by taxa

Taxa	Positive outcome (%)	Negative outcome (%)
Fungi		
Plant	0.00	16.67
Invertebrates	28.57	16.67
Fish	42.86	50.00
Amphibians	0.00	0.00
Reptiles	0.00	0.00
Birds	0.00	0.00
Mammals	28.57	16.67

Secondly, (M)PAs alone are most likely insufficient to avert the current extinction crisis (Mora and Sale 2011). (M)PAs are just part of the solution to the biodiversity crisis. They are spatially restricted (Noss et al. 2012; Wilson 2016) and do not yet cover essential biodiversity hotspots (Protected Planet 2020). Moreover, existing (M)PAs reduce but do not generally avoid biodiversity degradation and loss within their boundaries (Geldmann et al. 2020). However useful (M)PAs might be, it seems evident that if the only natural sites specifically designated to conserve biodiversity cannot avoid biodiversity loss, (M)PAs must be complemented with other effective policies and tools that help retain as much biological diversity as possible such as OECMs (IUCN 2020), gene banks, seed banks, botanic gardens or zoos. Such tools are intended to reduce biodiversity loss in the face of increasing human development pressures (WWF 2020). They can thus be regarded as 'defensive' conservation strategies of globally besieged biodiversity. If the current alarming rates of biodiversity loss are to be reversed, (M)PAs and other 'defensive' conservation tools will need to be complemented with 'offensive' actions aimed at increasing both the quantity and quality of habitats through widespread ecological restoration in key places for biodiversity (IPBES 2019).

Thirdly, (M)PA effectiveness is most likely determined by a number of contextual variables, notably legal stringency, managerial effort, biophysical characteristics and inner and outer pressures (Gaston et al. 2008; Mora and Sale 2011). (M)PAs are intended to abate some important pressures to biodiversity like deforestation or poaching. It was found that well-funded (M)PAs are much more effective at reducing pressures to biodiversity and thus at benefiting threatened species (Kearney et al. 2020). However, it is unrealistic to think that (M)PAs will prevent all existing pressures on their own, even if they are well managed and funded (Rodríguez-Rodríguez 2012; Kearney et al. 2020). Thus, whereas (M)PA effectiveness can most likely be enhanced through legal and managerial means, it remains highly site-specific and cannot be generalised (Mora and Sale 2011; Coetzee et al. 2014).

Fourthly, inefficient PAs should be clearly identified, assessed on the causes of their ineffectiveness and made effective or possibly degazzeted, so that scarce conservation resources could be devoted to enhancing conservation elsewhere where they could be most helpful (Fuller et al. 2010).

Finally, sound, consistent and regular biodiversity monitoring and evaluation schemes should be put in place in all (M)PAs to make sure their current effectiveness is maintained or, where needed, enhanced, prioritising threatened biodiversity. Well-designed biodiversity monitoring schemes are the basis of reliable (M)PA evaluations (Hockings et al. 2006; Addison 2011). However, they are complex, costly and time-consuming. Thus, they are seldom implemented, even in managed (M)PAs (Addison et al. 2015; Collen et al. 2016). Adequate monitoring has been an (M)PA managerial and research challenge for a long time (Hockings et al. 2000), and it remains so in most settings today (Addison et al. 2015; Protected Planet 2020), despite the use that global databases can provide (IUCN 2021b; WWF 2021). As a result, existing evidence is largely based on case studies using different species, monitoring techniques and conservation indicators across countries, regions and realms (Collen et al. 2016) (Fig. 6.5).

To make things more complex, many studies apparently addressing (M)PAs' effectiveness in conserving species (which should be studied using population trends from carefully collected monitoring data) are in fact representation studies of species in (M)PAs or (M)PA networks (Vieilledent et al. 2013; Santini et al. 2016). Representation is just a basic proxy for effective conservation. Thus, these studies are of some use to ascertain potential (M)PA effectiveness, especially if biodiversity data is spatially overlapped with legal designations and management data (Rodríguez-Rodríguez and Martínez-Vega 2018a). However, they clearly lack a sound basis to claim real (M)PA 'effectiveness'. To further complicate matters,

Fig. 6.5 Sound biological monitoring is essential for protected area effectiveness assessment. (*Credit*: David Rodríguez)

potential easily measured proxies to (M)PA effectiveness such as management effectiveness metrics do not adequately reflect conservation outcomes, so automatic correspondence between the two cannot be assumed (Leverington et al. 2010a, b; Carranza et al. 2014a; Rodríguez-Rodríguez et al. 2015a; Pyhälä et al. 2019).

Given the widespread limitations in funding for basic managerial activities in most PAs (Fuller et al. 2010; Kearney et al. 2020), PA managers can resort to a number of cost-effective options for implementing essential biodiversity monitoring activities. One is setting up citizen-science monitoring programmes that will train local volunteers in monitoring biodiversity according to sound scientific protocols (Van Swaay et al. 2015; Chandler et al. 2017). Other is establishing long-lasting cooperation agreements with nearby universities and research centres to carry out biodiversity studies of interest to (M)PA managers, including monitoring, as part of the students' master's or PhD's theses or internships, such as the Research Partners Programme implemented by Parks Victoria in Australia (Parcs Victoria 2021) (Fig. 6.6).

Fig. 6.6 Citizen-science schemes and partnerships with research centres and universities can meet important monitoring needs in protected areas. (*Credit*: David Rodríguez)

6.1.2 What Indicators, Realms, Biodiversity Types, Regions and Countries Have Been Assessed Most?

Indicators related to LULC change (29.06%) and species' richness, abundance and density (28.21%) were the most frequently used to assess (M)PA effectiveness (Fig. 6.7).

The majority of studies referred to terrestrial PAs, with one quarter relating to MPAs and just 4% relating to freshwater PAs (Table 6.2). Despite the fact that MPA area notably exceeds that of PAs (Protected Planet 2020), (M)PA effectiveness assessment effort seems however balanced according to the distribution of multicellular species, with 80% of them dwelling terrestrial environments, 15% inhabiting seas and oceans and around 5% living in freshwater environments (Grosberg et al. 2012). Terrestrial PAs were likely most studied due to the greater cost-effectiveness of surveying terrestrial environments, especially ecosystems, using freely available data, remote sensing data and other spatially explicit secondary data (Duro et al. 2007; Vihervaaram et al. 2017). Marine biodiversity data collection is, in contrast, more costly and challenging, both area-wise and operational-wise, often requiring primary data from costly on board devices and/or *in situ* samplings by

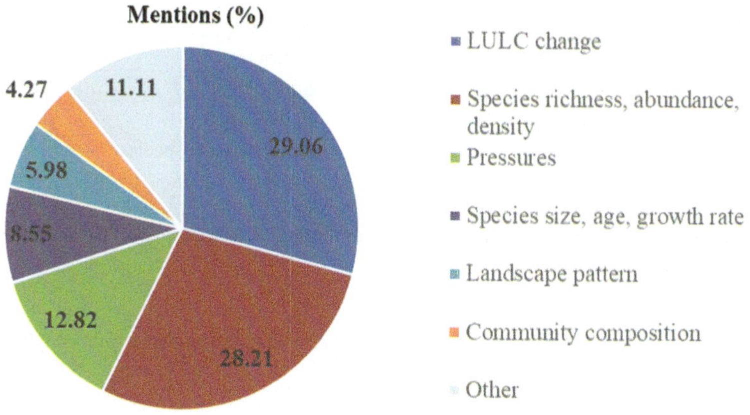

Fig. 6.7 Headline indicators used to assess (M)PA effectiveness

Table 6.2 Proportion of studies addressing different realms

Realm	Studies (%)
Terrestrial	61.84
Freshwater	3.95
Marine	25.00
Coastal	9.21

scuba divers (Hill et al. 2014; Bicknell et al. 2016; Egerton et al. 2018). Scarce data on freshwater biodiversity is concerning due to the greater global degree of threat of such biodiversity (Young et al. 2016; WWF 2020).

The type of biodiversity most widely studied were ecosystems, chiefly through LULC proxies (Fraser et al. 2009; Singh et al. 2014). Actually, some international organisations advocate the use of LULC to monitor pressures on biodiversity (European Environment Agency 2021; OECD 2021) and assess progress towards global programmes and agreements, such as REDD+ (Reducing Emissions from Deforestation and Forest Degradation; FAO 2021). In contrast, the effectiveness of (M)PAs in conserving genes was treated by just one study (1.3%; Table 6.3). Genetic biodiversity assessment is still a rarity in (M)PA effectiveness studies and broader conservation studies despite the evolutionary and conservation importance of this component of biodiversity (Hendry et al. 2010; Laikre 2010), even more so in the face of human-driven rapid environmental changes (Hoffmann et al. 2017).

A large bias in the assessed taxa was shown, with fungi and invertebrates (especially, insects) being severely underrepresented. In contrast, birds were notably less represented in our results than in previous biodiversity studies, where they tended to dominate alongside mammals (Barnes et al. 2016; Troudet et al. 2017). Most studies addressing (M)PA effectiveness in conserving taxa used fish (over 43%), whereas just 2.6% of all the reviewed studies included reptiles and amphibians, and none studied fungal, bacterial or Protista biodiversity (Table 6.4). Our results do not only

Table 6.3 Proportion of studies addressing different biodiversity types

Biodiversity	Studies (%)
Genes	1.32
Species	31.58
Ecosystems	51.32
Ecosystems & species	15.79

Table 6.4 Proportion of (1) all reviewed studies and (2) reviewed studies that addressed genes or species by taxa

Taxa	Studies (%)	Studies addressing genes or species[a] (%)
Bacteria	0.00	0.00
Protista	0.00	0.00
Fungi	0.00	0.00
Plant	6.58	13.51
Invertebrates	13.16	27.03
Fish	21.05	43.24
Amphibians	1.32	2.70
Reptiles	1.32	2.70
Birds	3.95	8.11
Mammals	10.53	21.62

[a]This column does not sum 100 because some studies included different taxa

highlight a very marked bias in the assessed taxa but also suggest that a very substantial part of the world's biodiversity in (M)PAs has not yet been studied (Oliveira et al. 2017; Pyhälä et al. 2019), even in developed countries (Hill et al. 2014). Especially concerning are the lack of data on a number of relevant taxa, including whole kingdoms such as Fungi or Protista, as well as the paucity of information on globally threatened taxa like amphibians, or the severe underrepresentation of prevailing taxa like invertebrates (Young et al. 2016) (Fig. 6.8).

We found a good geographical balance among the reviewed studies (Table 6.5), in contrast to reported geographical biases in biodiversity studies (Trimble and van Aarde 2012). Oceania was, however, substantially less studied than the other continents, as shown in previous works (Coetzee et al. 2014).

Studies from 36 countries were assessed. Of them, four countries concentrated over one third of the studies on PA effectiveness: China, Spain, Brazil and Tanzania. Two of them (China and Spain) accounted for more than one fifth of the total studies in the 2010–2019 decade (Table 6.6). These results highlight both, an important gap and bias in (M)PA effectiveness evaluation studies in the last decade. Some authors state the importance of intensifying cooperation to more effectively address societal biodiversity information needs (Meyer et al. 2015) (Table 6.7).

Fig. 6.8 Invertebrates are largely underrepresented in biodiversity studies. (*Credit*: David Rodríguez)

Table 6.5 Regional scope of the studies, in percentage

Region	Studies (%)
Africa	18.42
America	23.68
Asia	26.32
Europe	27.63
Oceania	3.95

Half of the studies assessed 'reserves', whereas 25% assessed multiple-use (M) PAs and 25% assessed both types of (M)PAs. The number of (M)PAs studied was 1,221,798, ranging from 1 to 1,216,275 per study.

6.1.3 *What Are the Most Relevant Pressures to PAs?*

Two realm-specific pressures outstood from our analysis: deforestation for PAs and fishing for MPAs. Actually, pressures leading to negative LULC changes from logging, urbanisation, agriculture and other uses stood for over one third of all mentioned pressures. In turn, fishing and overharvesting-related pressures accounted for over 16% of all mentioned pressures, whereas hunting and poaching represented just 3.6%. These results suggest that the major pressures to terrestrial and freshwater biodiversity come from habitat destructive practices, whereas the direct extraction of organisms is the major pressure to marine biodiversity, far ahead of other more mediatic pressures like climate change, as shown in previous studies (IPBES 2019; WWF 2020). Both pressures were considered not to have been substantially

Table 6.6 Countries represented in the studies

Country	Studies (%)
China	10.53
Spain	10.53
Brazil	7.89
Tanzania	5.26
USA	3.95
India	3.95
Italy	3.95
Philippines	3.95
Australia	2.63
Burkina Faso	2.63
Cameroon	2.63
Ecuador	2.63
Mexico	2.63
Russia	2.63
South Africa	2.63
Taiwan	2.63
UK	2.63
Canada	1.32
Chile	1.32
Colombia	1.32
Costa Rica	1.32
Egypt	1.32
France	1.32
Germany	1.32
Guatemala, Honduras, El Salvador	1.32
Hungary	1.32
Iceland	1.32
Indonesia	1.32
Ireland	1.32
Madagascar	1.32
Nepal	1.32
Nigeria	1.32
Micronesia Fed.	1.32
Portugal	1.32
Thailand	1.32
Vietnam	1.32

Table 6.7 Degree of development of the countries included in the studies, according to the Human Development Index

Development	Studies (%)
Very high	36.84
High	39.47
Medium	10.53
Low	10.53

Fig. 6.9 LULC changes are the main global pressure on terrestrial biodiversity. (*Credit*: David Rodríguez)

reduced globally in the last decade, thus largely failing to meet related international targets, notably Aichi Targets 5[2] and 6[3] (CBD 2020) (Fig. 6.9).

LULC changes were also considered a major combined pressure source for coastal PAs and even MPAs (Loiseau et al. 2021). A recent study found that maritime traffic and fishing were the most ubiquitous pressures in European MPAs (present in 66% and 32% of European MPAs, respectively), ahead of seabed habitat destructive pressures such as dredging, submarine installations or mining (Perry

[2] By 2020, the rate of loss of all natural habitats, including forests, is at least halved and where feasible brought close to zero, and degradation and fragmentation is significantly reduced.

[3] By 2020 all fish and invertebrate stocks and aquatic plants are managed and harvested sustainably, legally and applying ecosystem-based approaches, so that overfishing is avoided, recovery plans and measures are in place for all depleted species, fisheries have no significant adverse impacts on threatened species and vulnerable ecosystems and the impacts of fisheries on stocks, species and ecosystems are within safe ecological limits.

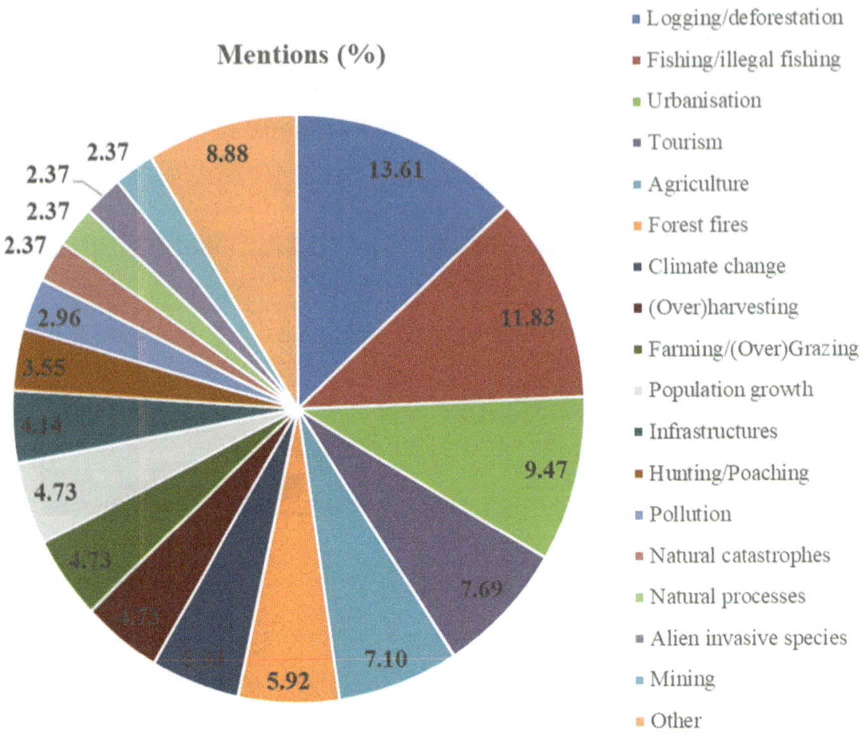

Fig. 6.10 Stated importance of pressures to (M)PAs according to their number of mentions

et al. 2020). Thus, key common components of (M)PA effectiveness (which should translate into conservation, managerial and evaluation objectives) will be the degree to which (M)PAs impede negative habitat transformations and the (over)harvesting of living organisms (Mora and Sale 2011) (Fig. 6.10).

In their global study, Bowler et al. (2020) found that, unlike in the marine realm, terrestrial pressures to biodiversity tended to considerably overlap spatially. In turn, other studies found that cumulative pressures to the marine environment (Halpern et al. 2008) and MPAs (Rodríguez-Rodríguez et al. 2015b) were greater nearshore than offshore. Both results suggest that achieving effective conservation in terrestrial and coastal (M)PAs is likely more challenging than for offshore MPAs, despite the fact that insufficient enforcement of many offshore MPAs, especially large ones, jeopardises their effectiveness, even if less pressured (Collins et al. 2021).

Pressures on biodiversity are on the rise despite increasing protection efforts (IPBES 2019), resulting in steady biodiversity declines (Butchart et al. 2010; WWF 2020). Geldmann et al. (2020) found that human pressures to global PAs had increased between 1995 and 2010, with the largest increases in south-east Asia and tropical Africa and the least increase in Oceania and North America. Increasing human pressures and limited (M)PA effectiveness make it essential to ascertain and streamline the factors that contribute to (M)PA performance (Fig. 6.11).

Fig. 6.11 Deforestation remains unabated in the biodiversity-rich tropics. Burnt dry forest in Madagascar. (*Credit*: David Rodríguez)

6.1.4 What Factors Contribute to the Success or Failure of PAs at Conserving Biodiversity?

Three major factors were considered by the studies' authors to contribute most to (M)PA effectiveness: site's *Management* (34.5%), *Legal protection*/conservation regulations (25.4%) and *Policy/politics* (16.4%), as shown in Fig. 6.12. Similar results were found in a previous SLR in which regulations and natural characteristics (e.g. remoteness) had had the greatest positive effect on habitat degradation, whereas managerial actions and regulations had had the greatest positive contribution to species' conservation (Geldmann et al. 2013).

(M)PA management has long been considered the main factor to (M)PA effectiveness (Hockings et al. 2000; Dudley 2008). Quite sensibly, managerial activities aimed at abating pressures such as surveillance and/or conserving biodiversity (e.g. reintroductions) positively influence conservation outcomes (Edgar et al. 2014; Kearney et al. 2020), even if they may not always suffice to stop or reverse the degradation of biodiversity (Mora and Sale 2011; Coad et al. 2015).

Legal aspects such as regulation stringency or zoning ranked second to (M)PA effectiveness. Even though legal factors have long been neglected or downplayed

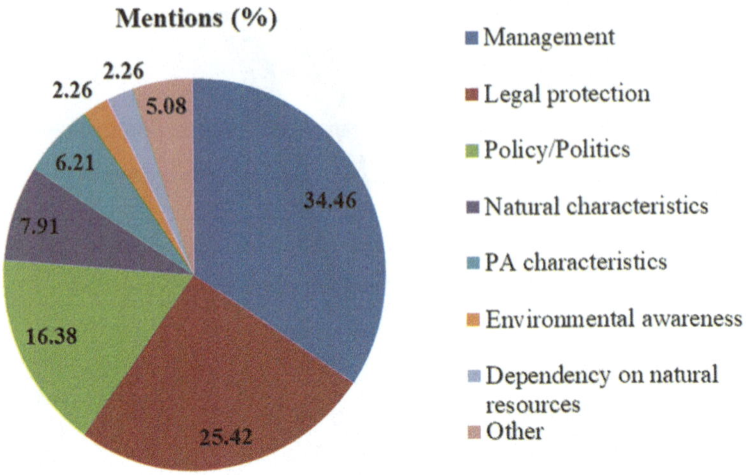

Fig. 6.12 Protected area effectiveness factors from the systematic literature review

through the euphemism 'paper park' (Rodríguez and Rodríguez-Clark 2001; Perry et al. 2020), they were shown to substantially contribute to (M)PA effectiveness, especially in contexts of reasonably well-performing institutions (Giakoumi et al. 2018; Rodríguez-Rodríguez and Martínez-Vega 2018b) (Fig. 6.13).

Another often downplayed factor of (M)PA effectiveness is politics, including the development of sustainable policies, political stability and political will. Nevertheless, it is politicians who often decide on essential legal and managerial aspects of (M)PAs such as designation itself, boundaries, legal stringency, legal overlap, degazzeting, zoning, co-management or funding (Gardner et al. 2018). It was found that deforestation rates in PAs substantially increased in Madagascar after the power void following the 2009 coup (Eklund et al. 2019). Other authors have pointed out the importance of other factors for the effectiveness of (M)PAs such as remoteness (Edgar et al. 2014), stakeholder engagement (Giakoumi et al. 2018), slope, road density, human influence and human development index (Geldmann et al. 2020). Later on in this chapter, we will summarise the effectiveness factors identified by the authors of the articles of four selected success stories. Conversely, we will present the explanatory factors that, in the opinion of their authors, may help to understand the lack of effectiveness of another four case studies.

In the data from our SLR, 'biodiversity outcomes' were moderately and inversely correlated with 'time range' ($r_{(52)} = -0.45$; $p = 0.001$), which means that positive outcomes came mostly from one-off studies whereas negative outcomes chiefly resulted from broad time range studies. This may suggest that the negative impact of globally mounting pressures in and around (M)PAs (IPBES 2019) might overcome the opposing impact of conservation in some (M)PAs and thus tend to reduce PA effectiveness with time (Geldmann et al. 2015, 2020).

Fig. 6.13 Vatnajökulspjódgardur National Park, Iceland. (*Credit*: Javier Martínez)

'Biodiversity outcomes' were also weakly correlated with 'Indicator trend' ($r_{(35)} = 0.39$; $p = 0.018$), so the better the outcomes, the better the trend of the assessed indicators tended to be, and vice versa. Kruskal-Wallis tests ($\alpha = 0.05$) returned no statistically significant differences in 'Biodiversity outcomes' (dependent variable), by 'PA type' ($p = 0.650$), 'Realm' ($p = 0.597$), 'Development' ($p = 0.341$), 'Country' ($p = 0.254$) or 'Region' ($p = 0.084$), and was on the verge of significance for 'Method' ($p = 0.065$) and 'Biodiversity type' ($p = 0.066$). 'Biodiversity outcomes' were however significantly different according to 'Time range' ($p = 0.003$), as previously mentioned, 'Indicator trend' ($p = 0.004$), and the 'Indicators used' ($p = 0.016$). These results further suggest that (M)PA effectiveness is highly site-specific, influenced by the assessment period, indicator-related and that it cannot usually be generalised.

'Indicator trend' was not significantly different across 'Realms' ($p = 0.971$), 'Country' ($p = 0.701$), 'Methods' ($p = 0.631$), 'Indicators used' ($p = 0.488$), 'Biodiversity type' ($p = 0.480$), 'Time range' ($p = 0.412$), 'Sampling method' ($p = 0.396$), 'Region' ($p = 0.373$), 'Development' ($p = 0.347$) or 'PA type' ($p = 0.152$), also suggesting high specificity.

6.2 Methodological Approaches to Assessing PA Effectiveness

In this section, we will summarise the results from the SLR that respond to the following methodological research question: *what are the methods most widely used to assess PA effectiveness?*

Over two thirds of the studies used moderately reliable, incomplete semi-experimental *research designs*. Given that (M)PA samples are not randomly distributed across the landscape/seascape, but rather subject to ecological and socio-economic related biases, sheer experimental research designs with randomised cases and controls cannot be applied to (M)PA effectiveness evaluations (Andam et al. 2008). The most internally valid studies using complete semi-experimental BACI designs (Andam et al. 2008; Addison 2011) accounted for less than one third of the total (Table 6.8).

Twenty-nine per cent of the studies used complete, semi-experimental research designs or methods[4] (BACI or BACI+), and 74% of them used incomplete, semi-experimental methods (deconstructed BACI designs). Just 1.3% of the studies used the most reliable MPBACI designs, whereas almost 20% used little reliable After-Impact designs (Table 6.9).

The *research designs* (BACI-type designs; Fig. 6.14) used to assess (M)PA effectiveness can differ in their estimations of effectiveness. Results from studies using automatically created buffer zones around (M)PAs were found to purport greater (M)PA effectiveness than studies selecting 'matched' counterfactuals of similar biophysical characteristics to (M)PAs, thus overestimating (M)PA effectiveness. Part of such effectiveness was found to be unrelated to protection measures and should

Table 6.8 Research designs used in the studies, in percentage. Overall percentages do not sum 100 because some studies used different designs for different indicators

Research design	Studies (%)
Complete semi-experimental	31.58
Incomplete semi-experimental	69.74
Mixed	2.63

Table 6.9 Assessment methods used in the studies, in percentage. Overall percentages do not sum 100 because some studies used different designs for different indicators

Assessment method	Studies (%)
MPBACI[a]	1.32
MBACI/PBACI	9.21
BACI	18.42
(X)ACI/BAI	15.79
ACI/BAI	38.16
AI	19.74

[a]Multiple-Paired-Before-After-Control-Impact

[4]Figures do not exactly match those from Table X due to the fact that some studies used different methods for different indicators (mixed research designs).

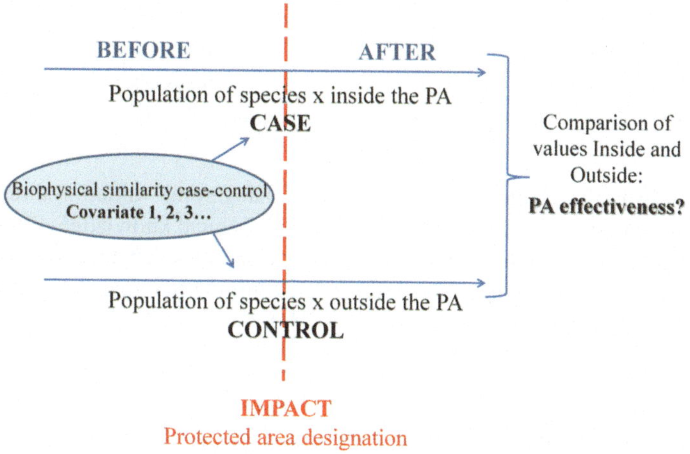

Fig. 6.14 Conceptual outline of BACI designs

Table 6.10 Assessment techniques used in the studies, by realm (in percentage). Column percentages do not sum 100 because some studies used different techniques

Technique	Total studies (%)	Terrestrial (%)	Marine (%)	Coastal (%)	Freshwater (%)
In situ sampling	53.95	43.90	43.90	7.32	4.88
Remote sensing	38.16	82.76	3.45	13.79	
GIS	15.79	83.33	8.33	8.33	
Interview	10.53	62.5	12.5	25	
Literature review	10.53	87.5			12.5
Aerial photographs	6.58	80.00	20.00		

instead be attributed to intrinsic biophysical characteristics that 'naturally' protect those sites from human-made pressures, such as remoteness, altitude and steepness. Thus, such factors should be taken into account as covariates for providing a more accurate picture of (M)PA effectiveness (Andam et al. 2008; Spracklen et al. 2015).

The main biodiversity assessment *techniques* used included *in situ sampling* (mostly for species' indicators) and *remote sensing* (chiefly for ecosystems' indicators). *In situ sampling* clearly prevailed in the marine realm, whereas a combination of *literature review* and remote analysis techniques (remote sensing, GIS and aerial photographs) was mostly used in the terrestrial realm (Table 6.10). Such figures align with those found in a previous literature review on PA effectiveness until 2010 by Geldmann et al. (2013).

Only 5.26% of all the studies ($n = 4$) assessed censuses of (M)PAs or complete, national (M)PA networks, all of them in Europe, including a pan-European assessment (Hellwig et al. 2019). In turn, 94.74% of the studies used samples of (M)PAs, ranging from 1 to 299.

In over 60% of the studies, the assessment spanned more than one decade (Table 6.11).

Table 6.11 Time range of the studies, in percentage

Time range	Studies (%)
Broad	60.53
Moderate	17.11
One-off	22.37

Just 13.16% of the studies were evaluations using some valuation threshold for the assessed indicators, which suggests limited ecological knowledge, little concretion of conservation or management objectives and/or substantial decoupling of research from (M)PA management, all of them common hindrances to (M)PA evaluation (Hockings et al. 2006; Addison et al. 2015).

6.3 Successes and Failures: Lessons Learned

In this section, we will delve on the results from the SLR that respond to the following thematic research questions:

1. Are there any positive and negative case studies to draw lessons from?
2. What factors contribute to the success or failure of PAs at conserving biodiversity?

Next, we present a diversity of case studies from our SLR as examples of success and failure on (M)PA effectiveness (Fig. 6.15), and expand on the general factors for PA success or failure at conserving biodiversity as lessons learned for key stakeholders such as managers, decision-makers and researchers.

6.3.1 Successful Cases

6.3.1.1 Case Study 1: Cabo Pulmo National Park, Mexico

6.3.1.1.1 Description of the Site

Cabo Pulmo National Park (CPNP) is located in southern Baja California, Mexico. It was designated a national park in 1995. In addition, it has been a Ramsar site since 2008, and a Natural World Heritage Site by UNESCO since 2013. Finally, it is included in the Mexican Programme for the Conservation of Species at Risk, PROCER (CONANP 2018). It covers an area of 7,111 hectares (Fig. 6.16), 35% of which is designated as a 'no-take' marine reserve. The local community is actively involved in the management of the PA and enforces a no-fishing policy throughout the reserve.

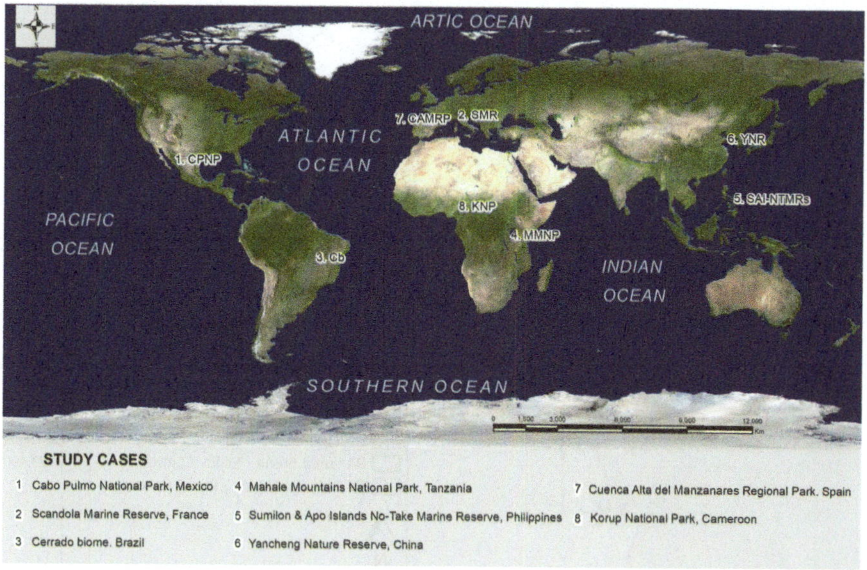

Fig. 6.15 Location of selected case studies. (Cartographic base: NASA. https://modis.gsfc.nasa.gov/data/)

Cabo Pulmo is the oldest known coral reef in the American Pacific, estimated to be 25,000 years old (Squires 1959). CPNP is composed of a mixture of habitats located on basaltic rocky reefs (Fig. 6.17). Between the rocky reefs are seafloor habitats.

6.3.1.1.2 Protected Biodiversity and Threats

In this national park, 226 reef species have been inventoried out of the 875 ichthyological species listed in the Gulf of California, as well as 11 of the 14 known species of hermatypic coral in this area (e.g. *Pavona gigantea* and *Pavona clivosa*). Multiple commercial marine species are concentrated here. Among those fished within the reef are grouper, garropa, cabrilla, snapper and parrotfish. Among the molluscs, the mother of pearl stands out. It is used for pearl extraction with commercial purposes.

Cabo Pulmo is home to species considered to be under some category of protection according to the Official Mexican Standard NOM-059-1994 (endangered, threatened and subject to special protection). Among the most outstanding ones are five species of sea turtles (*Caretta caretta*, *Chelonia agassizi*, *Dermochelys coriacea*, *Eretmochelys imbricata* and *Lepidochelys olivacea*) that are 'in danger of extinction' on a global scale. It also harbours different species of sharks (bull sharks, tiger sharks, hammerheads, blacktip and great white sharks), dolphins (*Tursiops truncatus*, *Stenella longirostris*, *Steno bredanensis*), whales (*Megaptera novaeangliae*) and fin whales (*Balaenoptera physalus*, *Balaenoptera edeni*).

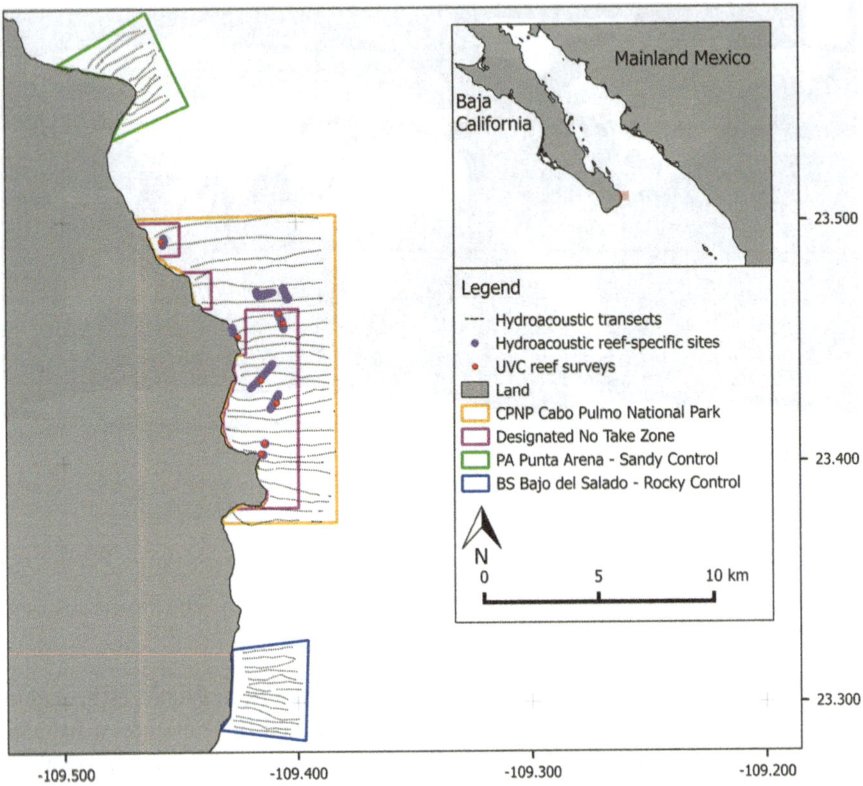

Fig. 6.16 Location of the no-take marine reserve (NTMR) within the Cabo Pulmo National Park (Mexico) and its outer control zones. Hydroacoustic transects and the Underwater Visual Censuses (UVC) are shown. (*Credit*: Egerton et al. (2018))

In the late 1990s and before the declaration of the national park and the establishment of the reserve zones, the main threats were commercial fishing and recreational activities linked to diving. Once both activities were regulated, the main concern was the tourism project called Cabo Cortés, which planned to build 30,000 hotel rooms, 2 golf courses and a marina with a capacity for 490 boats. Fortunately, the project was cancelled in 2012 because of pressure from local communities, environmental groups and national and international institutions responsible for environmental conservation.

6.3.1.1.3 National Park Effectiveness Assessment

To assess the effectiveness of the national park, Egerton et al. (2018) used an ACI-type (After Control-Impact) Incomplete Semi-Experimental (ISE) approach. They were unable to apply a Before-After Control-Impact (BACI) approach to the CPNP because no hydroacoustic records existed prior to its designation. To assess the

Fig. 6.17 Cabo Pulmo Beach. (*Credit*: John Markos O'Neill. https://commons.wikimedia.org/wiki/File:Cabo_Pulmo.jpg)

effects of protection against fishing, they compared fish density, fish biomass and mean fish size within the national park and in two control zones outside the NP. One of them, Punta Arena, is representative of sandy habitats. The other, Bajo del Salado, is a rocky reef complex. Both habitats are present in the NP. They carried out various hydroacoustic transects in March 2015.

6.3.1.1.4 Results

Fourteen years after the creation of the park, fish density, biomass and size were significantly higher inside the CPNP than in the control areas outside the national park. When Punta Arena and Bajo del Salado were examined together as a group of control areas and compared to CPNP, the biomass values in the national park were 273% higher and significantly different (Table 6.12).

Among the effectiveness factors, it is worth highlighting the effective management of the national park, the limitation of fishing in the reserve areas and the involvement of local communities in management, as well as their environmental awareness.

Table 6.12 Results of indicators measured in Cabo Pulmo National Park and surrounding control areas

Site	Mean density of fish/ha	Mean fish size (cm)
Cabo Pulmo National Park	447	11.40
Punta Arena (Sandy control)	130	6.02
Bajo del Salado (Rocky control)	99	9.50

Source: Egerton et al. (2018)

6.3.1.2 Case Study 2: Scandola Marine Reserve, France

6.3.1.2.1 Description of the Site

The Regional Natural Park of Corsica is a mixed marine and terrestrial protected area. It contains the Scandola Marine Reserve (590 ha), designated in 1975. It is one of the oldest MPAs in the Mediterranean. Its main objectives are the conservation of marine biodiversity and restocking of commercially valuable species. Moreover, since 1983 the marine reserve has been inscribed on the UNESCO World Heritage List.

It includes an integral reserve (IR) or integral protection zone (72 ha; see Fig. 6.18). Fishing, diving and anchoring of boats are prohibited. Most of the reserve has a partial protection status and is considered a buffer zone (BZ, 518 ha). In these zones, low-impact fishing activities are allowed, using traditional methods and small boats. In the surrounding unprotected areas (UP), professional, recreational and spearfishing is allowed.

6.3.1.2.2 Protected Biodiversity and Threats

On the one hand, the Scandola Marine Reserve protects coastal ecosystems and especially rocky reefs (Fig. 6.19) and, on the other hand, the fish communities of the *Posidonia oceanica* seagrass beds. Its waters are home to 450 species of algae and 125 species of invertebrates.

This case study (Harmelin-Vivien et al. 2015) focuses on the brown meagre (*Sciaena umbra*), a species listed as vulnerable in the Mediterranean according to the IUCN Red List of Threatened Species (Bizsel et al. 2011). It is a commercial species in the Mediterranean basin, particularly vulnerable to fishing pressure and, in particular, to spearfishing. There is a worrying population decline, and low abundance persists in the surrounding unprotected areas. According to personal communications, some older professional fishermen in other Mediterranean ports consider this species to be extinct.

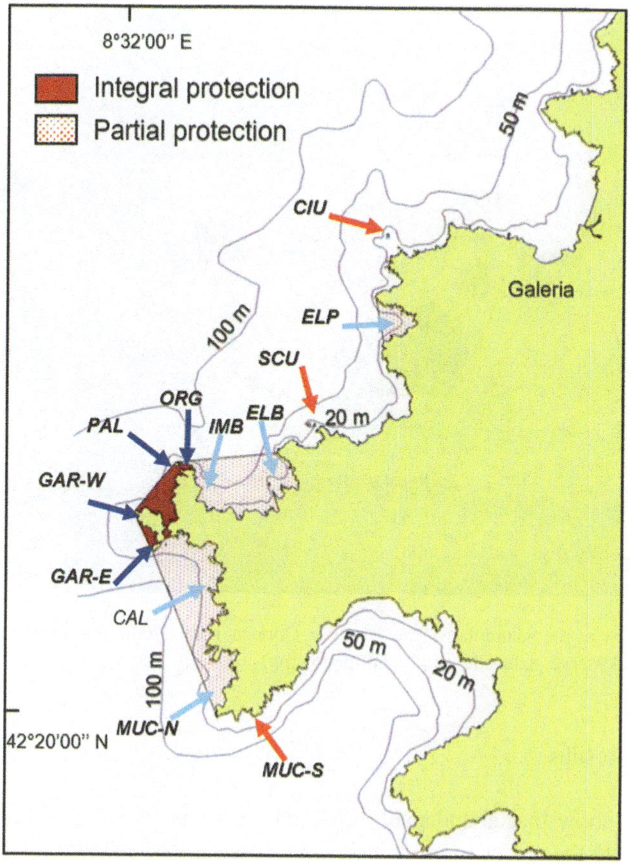

Fig. 6.18 Locations of sites investigated for Sciaena umbra survey at Scandola Marine Reserve (Corsica, France) according to protection level. Dark blue arrows, sites in no-take totally protected zone (IR); light blue arrows, sites in partially protected zones (BZ); red arrows, sites in unprotected fished zones (UP). The alphabetical codes of the sites are given in Harmelin-Vivien et al. (2015). (*Credit*: Professor Harmelin-Vivien)

6.3.1.2.3 Marine Reserve Effectiveness Assessment

To assess the effectiveness of the marine reserve, Harmelin-Vivien et al. (2015) used a MACI (Multiple-After Control-Impact) method within an ISE research design. The authors studied the density, abundance and biomass of *Sciaena umbra*, using visual censuses at different depths in the three zones (IR, BZ and UP). They conducted the surveys during the summer of 2012 along 532 transects at the 12 sites shown in Fig. 9.4, in similar habitats (*P. oceanica* seagrass meadows, rocky reefs or a mixture of these two habitats) located in the 3 zones (IR, BZ and UP). To understand the long-term effects of the marine reserve designation, they compared the data obtained in this study with a similar one carried out in 1983.

Fig. 6.19 View of the Scandola Marine Reserve, Corsica. (https://commons.wikimedia.org/wiki/File:R%C3%A9serve_naturelle_de_Scandola_(8).jpg)

6.3.1.2.4 Results

The results show that abundance, individual size and biomass of brown meagre increased with the degree of protection of the reserve. The abundance of larger individuals and the number of fish per shoal were significantly higher in IR than in UP. The abundance of this species was also higher in IR than in UP in 1983. The results demonstrate the long-term beneficial effect of this reserve on this vulnerable species.

Protection measures for brown meagre have been introduced in Corsica since January 2014. Spearfishing and some forms of recreational fishing have been banned. However, this moratorium is temporary and its renewal must be scientifically defended. The experts recommend maintaining long-term protective actions to promote the recovery of the populations of this species alongside sustainable regulation of professional and recreational fishing. On the other hand, it is recommended to promote periodic monitoring studies of vulnerable fish species inside and outside marine protected areas in order to estimate the time needed for full population recovery, to assess the carrying capacity of coastal ecosystems and to provide conservation managers with an assessment of the effects of their actions with regard to the effectiveness of fishing regulations.

6.3.1.3 Case Study 3: Cerrado Landscape, Brazil

6.3.1.3.1 Description of the Site

The 'Cerrado' landscape is a mosaic of savannahs, forests and grasslands (Fig. 6.20). Originally, it covered an area of 1,783,200 km^2 representing 23% of the area of the country. In this study, Carranza et al. (2014b) included 107 PAs and 62 indigenous lands (ILs). In addition, they divided the protected areas into 2 subsets (Fig. 6.21): 62 strictly protected areas (SPs) and 45 multiple-use protected areas (MUs). Both types of protection have different levels of land use restrictions and different management objectives (Dudley 2008). Broadly speaking, SPs correspond to IUCN categories I–IV, while MUs correspond to categories V–VI. The authors did not include private reserves because they were not mapped yet.

6.3.1.3.2 Protected Biodiversity and Threats

The Brazilian Cerrado has rich biodiversity, estimated at 10,000 plant species, of which 4,400 are endemic. In the savannah, there are about 800 species of trees and shrubs (Ratter et al. 1997). It is the habitat of 837 species of birds, 161 species of mammals, 120 species of reptiles and 150 species of amphibians. Among all

Fig. 6.20 View of Cerrado (Brazil. Jalapão State Park). (*Credit*: Rodrigo José Fernandes. https://commons.wikimedia.org/wiki/File:O_Cerrado_-_Jalap%C3%A3o_4.jpg)

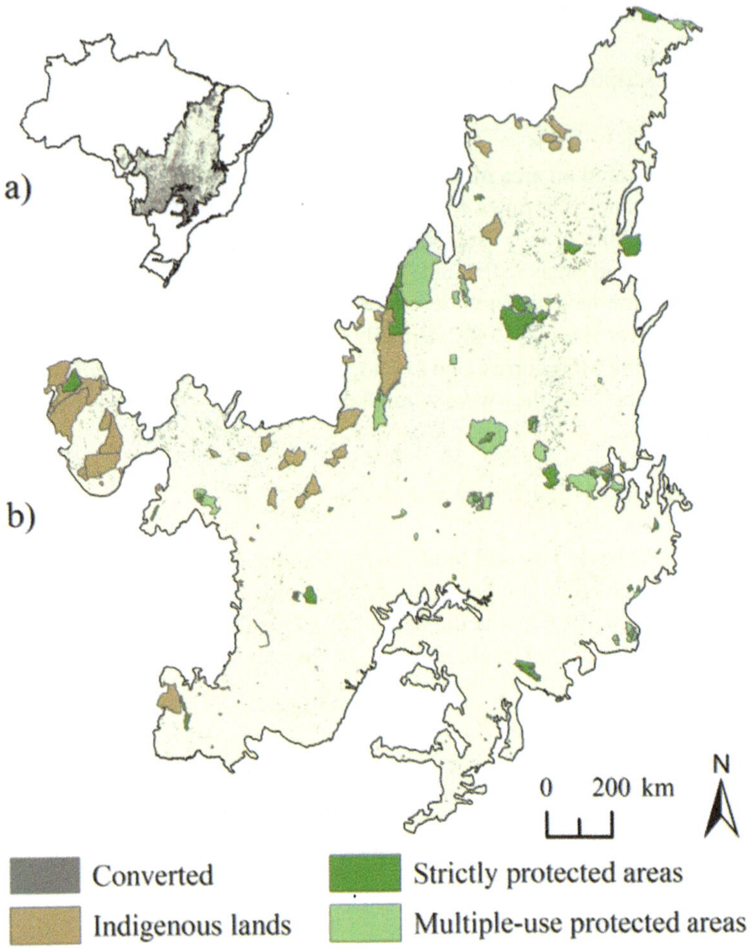

Fig. 6.21 Location of the Cerrado landscape in Brazil (**a**) and of the areas analysed in this case study (**b**). (*Credit*: Carranza et al. (2014b))

vertebrate species, 117 species are endemic. Globally, this landscape is a global biodiversity hotspot (Myers et al. 2000). About 22,000 km^2 is included within protected areas.

Since 1970, various processes have threatened the Cerrado. The literature has paid more attention to the transformation brought about by intensive agriculture and livestock.

According to Sano et al. (2010), extensive cultivated fields of soybeans, corn and rice and vast extensions of grasslands improved by man for cattle raising have replaced 40% of this landscape. Klink and Moreira (2002) consider that this ecological landscape is the main agricultural frontier of Brazil. Other economic activities have also left a deep mark on the territory. This is the case of the production of charcoal for the Brazilian steel industry. Finally, forest fires, landscape fragmentation and the spread of invasive species are also common pressures in the Cerrado. In summary, this landscape has suffered a higher deforestation rate than the Amazon rainforest. However, conservation efforts have been modest (Klink and Machado 2005).

6.3.1.3.3 PA Effectiveness Assessment

Firstly, the aim of this case study was to explore whether Cerrado habitat conversion was lower within PAs and ILs than in comparable unprotected areas. A second aim was to find out whether the levels of effectiveness vary between different types of PAs, depending on the legal stringency of the different types of protection.

To assess the effectiveness of PAs and ILs, Carranza et al. (2014b) used a MACI-type method within an ISE design. They used matching methods between the inner and outer zones of PAs. Using a Geographic Information System (GIS), they created a 250 × 250 m grid and analysed all the points that intersected with the selected PAs (1,137,972 points), with the ILs (1,202,068), with their corresponding 10 km buffers (1,471,059) and with other unprotected areas surrounding the buffers (8,129,688 points). They considered a moderate time series (2002–2009) of MODIS satellite images supported by a visual analysis of high-resolution images (Landsat, CBERS CCD). Additionally, they used seven independent variables (accessibility, travel time to the nearest city, rainfall, agricultural suitability, vegetation type, Cerrado physiognomies, elevation) as predictors of the dependent variable (Cerrado conversion). The predicted values of the logistic generalised linear model, ranging from 0 to 1, were used to match the treatment and control samples. In this way, they matched 85.7% of the original PA sample with the most similar controls, resulting in 975,661 matched pairs. For the IL sample, they matched 57.4% of the points, resulting in 690,425 pairs.

6.3.1.3.4 Results

Overall, the absolute effect of PAs in the Brazilian Cerrado has been positive. PAs have conserved savannah habitats better than their buffer zones and control sites. SPs have had a greater absolute effect than MUs. The ILs also had a positive absolute effect, even greater than that of the APs as a whole. There were no significant differences in habitat conversion between the 10 km buffer zones and the unprotected control sites surrounding the buffers. Thus, there was no evidence of a spillover.

Despite the above threats, the authors considered that the designation and management of PAs with complementary regulations (strict, multiple-use) and indigenous lands were effective factors that are contributing to slowing down the accelerated conversion of the Brazilian Cerrado. Finally, in addition to Cerrado conversion, Carranza et al. (2014b) recommended studying other threats such as habitat degradation (e.g. by frequent fires), fragmentation or spread of invasive species.

6.3.1.4 Case Study 4: Mahale Mountains National Park, Tanzania

6.3.1.4.1 Description of the Site

The Mahale Mountains National Park (MMNP) was established in 1985. In western Tanzania, it covers a total area of 1,613 km^2, mainly terrestrial. The Mahale Mountains run through the western part of the park from northwest to southeast. Mount Nkungwe is the highest point (2,462 m). In addition to the altitudinal effect, this orographic layout conditions the distribution of vegetation and the uneven distribution of rainfall. Rainfall is more abundant on the western slopes than on the eastern, leeward slopes. Tropical rainforests occupy the west of the national park. Riverine forests are distributed along the banks of the rivers that run steeply down the western slopes towards Lake Tanganyika, and along the eastern slopes (Fig. 6.22). To the east, a savannah ecosystem, known locally as 'miombo' forest, extends over a wide semi-arid piedmont. Approximately, it occupies 75% of the national park. Rainfall ranges between 1500 and 2500 mm annually.

Fig. 6.22 Landsat ETM+ satellite image of Mahale Mountains National Park along Lake Tanganyika, Tanzania. The park's terrestrial borders are outlined in yellow. (*Credit*: Jesse Allen, NASA Earth Observatory. https://earthobservatory.nasa.gov/images/6014/mahale-mountains-national-park)

The national park also has a 1.6-km-wide aquatic belt that stretches along the eastern shore of Lake Tanganyika, occupying 96 km^2 (Sweke et al. 2016). Its elongated and narrow layout suggests that it lies on one of the planet's major tectonic trenches: the Great Rift Valley. For that reason, it is the world's longest lake and the second deepest and second largest in water volume. In short, there is a marked altitudinal and longitudinal (west-east) gradient. The result is a varied mosaic of ecosystems that are the basis of the park's uniqueness and the rich biodiversity of its flora and fauna. In the terrestrial zone, three major ecozones can be distinguished: the tropical rainforest, the savannah and the 'miombo' woodland.

6.3.1.4.2 Protected Biodiversity and Threats

In accordance with the geographical diversity of the national park, its flora is varied. The miombo woodland is dominated by *Brachystegia* and *Julbernardia* species. In the montane forest, there are trees belonging to the genera *Podocarpus*, *Bersama*, *Nuxia congesta*, *Macaranga* and *Croton*. The total number of plant species recorded in the Mahale Mountains National Park is 1,174, although these are less than half of the estimated total number. It is estimated that there are about 120 species of terrestrial mammals in Mahale. Of these, 82 species have been sighted or studied in Mahale Mountains National Park (Tanzania National Parks 2021). The rainforest is dominated by chimpanzees (*Pan troglodytes schweinfurthii*), giant forest squirrel (*Protoxerus stangeri*), red-legged sun squirrel (*Heliosciurus rufobrachium*), brush-tailed porcupine (*Atherurus* sp.), Angolan black-and-white colobus monkeys (*Colobus angolensis*), bushy-tailed mongoose (*Bdeogale crassicauda*), banded mongoose (*Mungos mungo*), blue duiker (*Cephalophus monticola*) and Sharpe's grysbok (*Raphicerus sharpei*). The savannah is the habitat of lions, Grant's zebras and giraffes. The 'miombo' woodlands are home to roan antelope, sable antelope and Lichtenstein's antelope. In addition, more than 350 species of birds have been sighted. All endemic taxa at Mahale are classified as 'Vulnerable' in the ICBP/IUCN Red Data Book.

The lake contains at least 250 species of cichlid fish (*Cichlidae*) and another 150 species of non-cichlid fish, most of which inhabit the lake's shoreline, littoral and abyssal zones. In the internal waters, six species dominate: two species of 'Tanganyika sardines' and four predatory species of the genus *Lates*. Almost all Tanganyika cichlid species are endemic. This high degree of endemicity of most species also occurs with many of the lake's invertebrates, such as molluscs and crabs.

According to its protection status, fishing and other economic activities are prohibited in the MMNP. Only, under a special licence, fishing is allowed within 1.6 km of the shoreline. Despite this, fishing activities in the surrounding waters remain unregulated. There, illegal fishing is frequent. Furthermore, there are few studies on the ecological effectiveness of the MMNP and the loss of fish abundance and diversity in adjacent areas.

6.3.1.4.3 National Park Effectiveness Assessment

This case aimed to delve into the effect of PAs on freshwater environments, little studied until now. To assess the effectiveness of the national park, Sweke et al. (2016) used an ACI-type method (After Control-Impact) within an Incomplete Semi-Experimental (ISE) research design. They compared the abundance and diversity of fish at five sampling points inside the national park – where fishing is not allowed – versus four control areas outside the national park (Fig. 6.23).

Sampling of fish and environmental parameters was conducted at depths between 5 m and 10 m. Sampling was replicated 12 times at each site using gillnets set perpendicular to the shore. Total catch, species and wet weights of fish were recorded at each site. Fish biodiversity in protected zones and control zones was calculated as the total amount of species turnover between sampled sites in both areas.

6.3.1.4.4 Results

The authors recorded 518 individual fish of 57 different species. Fish weight abundance was five times higher in the MMNP than in the surrounding unprotected areas. Fish abundance and diversity were also higher in the national park than in the surrounding areas. Moreover, both variables decreased with distance from the park. These findings confirmed the effectiveness of this PA and its importance for the conservation of fish resources in Lake Tanganyika.

Among the effectiveness factors of the MMNP, it is worth noting the legal stringency assigned to this protected area, the regulatory prohibitions on fishing

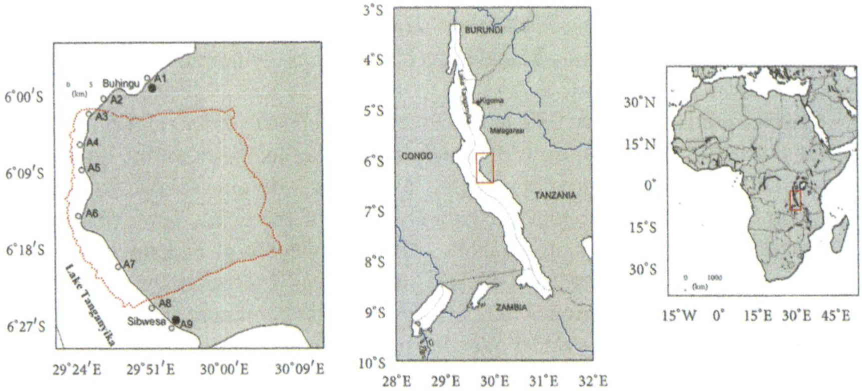

Fig. 6.23 Map of Lake Tanganyika showing study sites within protected area (MMNP, A3–A7) and in unprotected areas (Buhingu, A1–A2, and Sibwesa, A8–A9). The red polygon in the first frame shows sketch boundaries of MMNP (terrestrial and water strip in grey color and white color, resp.). The blue lines in the middle frame represent rivers. (*Credit*: Sweke et al. (2016))

activities and their enforcement. On the other hand, its geographical situation favours the conservation of its natural resources due to its scarce and difficult accessibility. On the contrary, in its surroundings, there is evidence that even the prohibition of beach seines is not respected. These could be some of the reasons for the differences in fish abundance and diversity between the national park and its unprotected surroundings.

The authors make some recommendations to improve the effectiveness of the PA and the biodiversity of the region. These include the extension of the national park and the establishment of new protected areas nearby. In addition, the application of adaptive management methods linked to the results of continuous ecological monitoring and warning signals is deemed desirable. These actions, together with the awareness of local populations, would favour the conservation of biodiversity and a sustainable use of fishing resources. The involvement of stakeholders in the co-management of resources would be highly recommended.

6.3.2 Lessons Learned from Success Stories

We can highlight some common factors from the cases whose results on biodiversity were positive. These characteristics were mentioned, directly or indirectly, by most of the authors of these case studies. We also summarise the common recommendations from those studies.

6.3.2.1 Effectiveness Factors

1. *Legal stringency*. All successful cases have been recorded in terrestrial or marine nature reserves and national parks (Carranza et al. 2014b; Harmelin-Vivien et al. 2015; Sweke et al. 2016; Egerton et al. 2018). As we know, all of them are characterised by applying strict restrictions to activities that could negatively impact their biodiversity. However, good results do not occur exclusively in these very demanding protection figures. On one occasion (Carranza et al. 2014b), there were also successful stories in multiple-use PAs or on indigenous lands. In this case, the involvement of the indigenous and/or local population was crucial, as we will see below.
2. *Community involvement*. In PAs where local communities and stakeholders are actively involved in the co-management of natural resources, successful outcomes seem more likely. This was a constant recommendation of marine and terrestrial PA researchers to their managers in order to improve their effectiveness.
3. *Continuous monitoring of threatened species and ecosystems* within PAs (temporal approach). In this sense, remote sensing and *in situ* sampling data exported to GIS environments are very useful tools for recurrent assessments. The former was commonly used for the assessment of indicators related to land use, habitats

or landscapes in terrestrial or mixed ecosystems. *In situ* surveys were widely used for the analysis of fish and invertebrate abundance and diversity in marine environments. Carranza et al. (2014b) and Harmelin-Vivien et al. (2015) used a moderate and broad temporal range, respectively, to monitor the ecosystems and species of the (M)PAs studied. The approach followed by Sweke et al. (2016) and by Egerton et al. (2018) is a one-off. However, in the case study of Tanzania, its authors point out the importance of temporal monitoring to ensure the success of biodiversity conservation. In the case of Cabo Pulmo, Egerton et al. (2018) mention the existence of a 14-year monitoring programme based on underwater visual censuses.

6.3.2.2 Recommendations

1. *Expansion of existing PAs and/or creation of complementary PAs* (Carranza et al. 2014b; Sweke et al. 2016). Among other benefits, they could facilitate connectivity between habitats, seed dispersal and movement between populations of target species (threatened, vulnerable, etc.). Some authors (Carranza et al. 2014b) point out the importance of studying other threats, such as habitat degradation due to forest fires, fragmentation or the spread of invasive species, which make it difficult to achieve biodiversity conservation objectives in and around PAs. Sweke et al. (2016) consider that the expansion of the MMNP should be carefully considered because it could redirect fishing pressure to unprotected areas and cause further management challenges there.
2. *Monitoring should also be done in the surroundings of the PAs*, considering that they are not 'islands' in the territory. This landscape spatial approach makes it possible to compare the processes and dynamics occurring within the PAs under study and those occurring outside them (Carranza et al. 2014b; Harmelin-Vivien et al. 2015; Sweke et al. 2016; Egerton et al. 2018).

6.3.3 Failure Cases

6.3.3.1 Case Study 5: Sumilon and Apo No-Take Marine Reserves, Philippines

6.3.3.1.1 Description of the Site

The study by Russ et al. (2017) was conducted on Sumilon and Apo Islands, located in the central Philippines (Fig. 6.24).

Apo and Sumilon Islands are two marine reserves in the Philippines. Sumilon is an island adjacent to Cebu Island. Apo is a small volcanic island in the Philippine archipelago (Fig. 6.25), close to the southeastern tip of Negros Island, surrounded by coastal coral reefs.

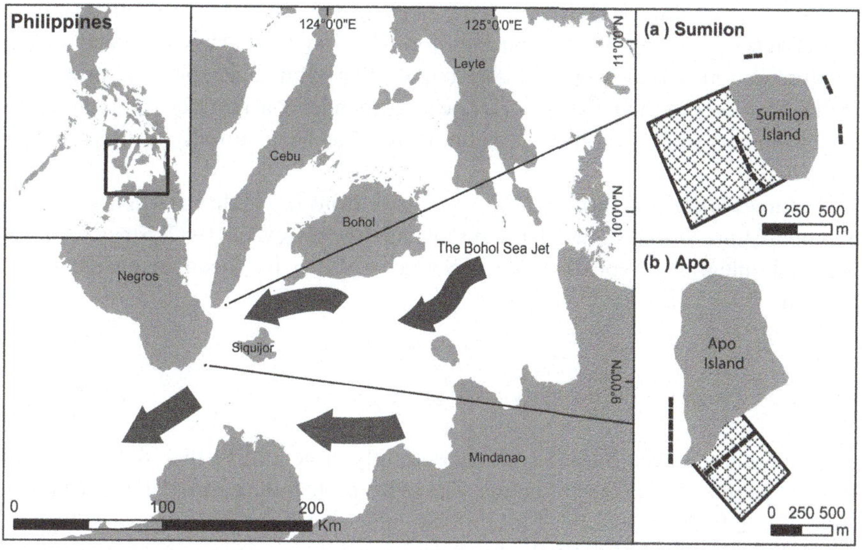

Fig. 6.24 Location of the study sites in the central Philippines. Inset (**a**): Sumilon Island. Inset (**b**): Apo Island. Cross-hatching indicates no-take marine reserve area. Black rectangles show approximate positions of permanent 50 m * 20 m replicate transects for fish and benthic surveys. (*Credit*: Russ et al. (2017). https://doi.org/10.1371/journal.pone.0188515.g001)

Fig. 6.25 View of Apo Island. (*Credit*: macoy.mejia. https://commons.wikimedia.org/wiki/File:Apo_Island_of_Apo_Reef_Natural_Park.jpg)

Thanks to Silliman University's research and a local government decree, a marine reserve was established on Sumilon Island in 1974. They designed awareness programmes to protect marine resources. For political reasons, full protection of the Sumilon Marine Reserve has been suspended on two occasions since its establishment, and on another occasion, fishing has been partially restricted (Russ et al. 2017).

The marine habitat surrounding Apo Island from the high water mark level to 500 m offshore was designated as a no-take marine reserve in 1982. Since then, the National Integrated Protected Areas Service (NIPA) has been in charge of its protection.

6.3.3.1.2 Protected Biodiversity and Threats

The Sumilon and Apo marine reserves currently protect more than 600 species of fish and nearly 400 species of corals. They also protect the entire coral reef habitat surrounding both islands. For decades, fishing has been the main source of income for its inhabitants. In the 1960s, fishermen began using unsustainable and illegal methods such as dynamite and 'muroami', a highly destructive fishing technique comprising an encircling net attached to pounding devices that are repeatedly launched on corals, smashing them. A decade later, nylon nets with a small mesh size and cyanide were introduced (Marten 2005). These methods damaged coral habitats and depleted even immature specimens, making the fisheries unsustainable (Alcala and Russ 2006).

In the early 1980s, the Philippine government banned fishing by these destructive methods. However, the large extent of coastal waters makes effective monitoring and enforcement difficult. In 1998, coral bleaching affected the islands of Sumilon and Apo. Today, both reserves are facing new threats related to climate change. In 2010, a local storm affected the Apo NTMR. In the following years, strong typhoons also damaged the reserve.

6.3.3.1.3 No-Take Marine Reserve Effectiveness Assessment

To assess the effectiveness of both marine reserves, Russ et al. (2017) used an ACI method within an ISE design. Each of the two islands has one 'fishing' and one 'NTMR protection' site. These 4 sites were monitored annually and consistently (26 times in 31 years) from 1983 to 2014 using Underwater Visual Censuses (UVC) by a single observer. The objective was to quantify the effects of marine reserve protection and benthic habitat changes on the density of coral reef fishes. In this case, Russ et al. (2017) studied five species of wrasse: two species of large-bodied (40–50 cm TL) *Hemigymnus* (*Hemigymnus melapterus* and *H. fasciatus*) that were vulnerable to fishing and three species of small-bodied (10–25 cm TL; *Thalassoma hardwickii*, *T. lunare* and *Cirrhilabrus* spp.) that were not targeted by fishing. Between 1983 and 1999, the benthic cover was estimated using linear transects. After 1999, benthic coverage and composition were obtained by visual observations along grids.

6.3.3.1.4 Results

The results of this study show that the density of labrid fishes is more influenced by changes in benthic habitats than by the protection provided by both marine reserves against fishing. They suggest that there are no significant differences in the density of target fish between the NTMRs and the outer fishing grounds. Tropical wrasses are at greater risk of extinction from environmental perturbations related to climate change than from fishing pressure. In summary, it seems prudent, according to the authors, to focus efforts on monitoring coral cover, coral composition and reef health. Although local communities were actively involved in the conservation of the reserves, the results were unsatisfactory.

6.3.3.2 Case Study 6: Yancheng Nature Reserve, China

6.3.3.2.1 Description of the Site

The site is a National Nature Reserve and a Ramsar site. In addition, together with the Dafeng Milu National Nature Reserve, it has been a UNESCO Biosphere Reserve since 1993. It represents the largest coastal mudflat nature reserve in China. It covers almost half a million hectares along a coastline of more than 120 kilometres. In ancient times, Yancheng was on the coastline, criss-crossed by a network of canals and protected by a system of dykes. The local population was engaged in fishing and salt trade. Due to repeated flooding, the Yellow River changed its course in the nineteenth century, abandoning its traditional course through the reserve. In the twentieth century, flood control and the introduction of irrigation systems led to the cultivation of cotton in the new land and the development of the textile industry in the province.

The Reserve has three management zones (Zang et al. 2017): the core zone, the buffer zone north and south of the core (NBZ, SBZ) and the experimental zone (NEZ, SEZ) (Fig. 6.26). In the core zone, wetland vegetation followed natural successional cycles. Currently, the site is owned by Yancheng City. The Reserve managers make decisions about the core zone, although local governments have some management rights over the buffer zones.

6.3.3.2.2 Protected Biodiversity and Threats

Following Zang et al. (2017), Yancheng National Nature Reserve (YNNR) is a relevant site in the East Asian-Australasian bird flyway. It is the habitat of 402 bird species (Fig. 6.27), 284 fish species, 31 mammal species, 26 amphibian and reptile

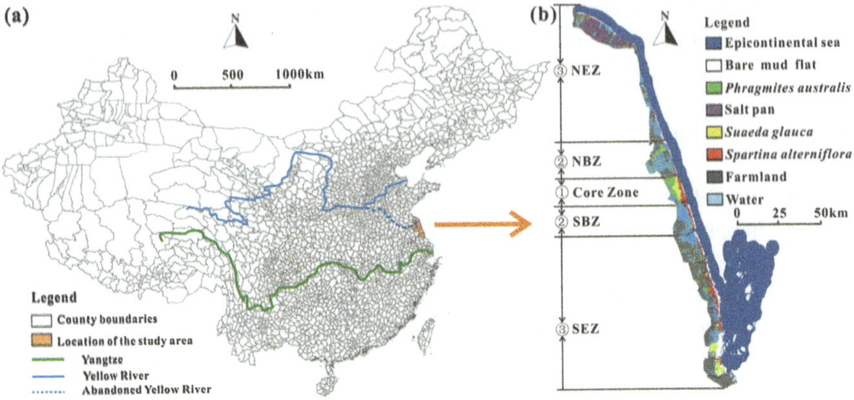

Fig. 6.26 Location of the Yancheng Nature Reserve (China) and distribution of dominant landscapes in the different management zones. (*Credit*: Zang et al. (2017). https://doi.org/10.1016/j.ecolind.2016.08.019)

Fig. 6.27 Red-crowned cranes at Yancheng National Nature Reserve in east China's Jiangsu province. (*Credit*: Xinhua. http://www.china.org.cn/china/2019-07/05/content_74956927.htm)

species and about 450 plant species. From sea to land, there are successive landscape bands of bare mudflats, *Spartina alterniflora* wetlands, *Suaeda glauca* wetlands and *Phragmites australis* wetlands.

Major threats to biodiversity include rapid urban and economic growth and high population density in Yancheng (Sun et al. 2016). In addition to land development, the construction of highways and railway lines has fragmented the landscape. Moreover, the experimental zone and buffer zone have been occupied by intensive agricultural land and aquaculture areas. This process of change has also contributed to the fragmentation of wetland landscapes. Along with human activities, climate change is also a factor that could increase the vulnerability of this coastal ecosystem. In particular, sea level rise is a potential threat.

6.3.3.2.3 Nature Reserve Effectiveness Assessment

To assess the effectiveness of this natural reserve, Zang et al. (2017) used a Before After Impact (BAI) method within an ISE research design. Through the analysis of land use land cover (LULC) changes, the objective of this study was to understand the relationship between YNNR landscape patterns and its ecological vulnerability and ecosystem services. The authors used the number of patches (NP), patch density (PD) and mean patch size (MPS) as indicators of the relationship with ecological vulnerability, and class area (CA) as an indicator of the relationship with the value of ecosystem services. The authors used a multi-temporal series of Landsat 5–8 images, between 1987 and 2013, supported by in situ reconnaissance work on the three natural wetland ecosystems.

6.3.3.2.4 Results

Among the processes of LULC changes, the expansion of agricultural land and the invasion of *S. alterniflora* were the most relevant (Xu et al. 2012). Other processes linked to natural factors, such as erosion and seawater deposition, also occurred. All of these caused changes in landscape patterns, ecological vulnerability and ecosystem services in the nature reserve. Fragmentation increased in all three wetland landscape types. Between 1987 and 1997, the value of ecosystem services provided by *S. glauca* and *P. australis* wetlands decreased, while that of *S. alterniflora* wetlands increased. However, in the period 1997–2013, the value of ecosystem services of all three wetland types decreased.

In summary, according to these indicators, the situation has worsened over the last few years. The authors recommended improving the effectiveness of the nature reserve by considering an adaptive management system that takes into account natural succession on the one hand, and flexible human intervention on the other hand, harmonising biodiversity conservation and economic development.

6.3.3.3 Case Study 7: Cuenca Alta del Manzanares Regional Park, Spain

6.3.3.3.1 Description of the Site

Following Schmitz et al. (2012), the Cuenca Alta del Manzanares Regional Park (CAMRP) has an area of 52,796 ha and a socio-economic peripheral zone (SIZ) of 39,204 ha that was mostly unprotected during the study time range. In this case, the study area comprises 25 municipalities, 13 within the park and 12 in the SIZ (Fig. 6.28). The Regional Park was designated by the Autonomous Region of Madrid in 1985. It has a multiple-use status equivalent to IUCN management category VI (Dudley 2008). In 1992 it was designated a biosphere reserve in recognition of the interaction between biodiversity conservation and human development. This area also forms part of the Natura 2000 Network, under the figures of Special Protection Area (SPA) and Special Area of Conservation (SAC). CAMRP surrounds Monte del Pardo, a Natura 2000 site of high ecological value. Finally, it is noteworthy that the Regional Park connects the city of Madrid with the Sierra de Guadarrama National Park, designated in 2013.

Since 1987, the CAMRP has had a management plan. It was reviewed in 1995 (Comunidad de Madrid 1997). According to it, the park is divided into seven

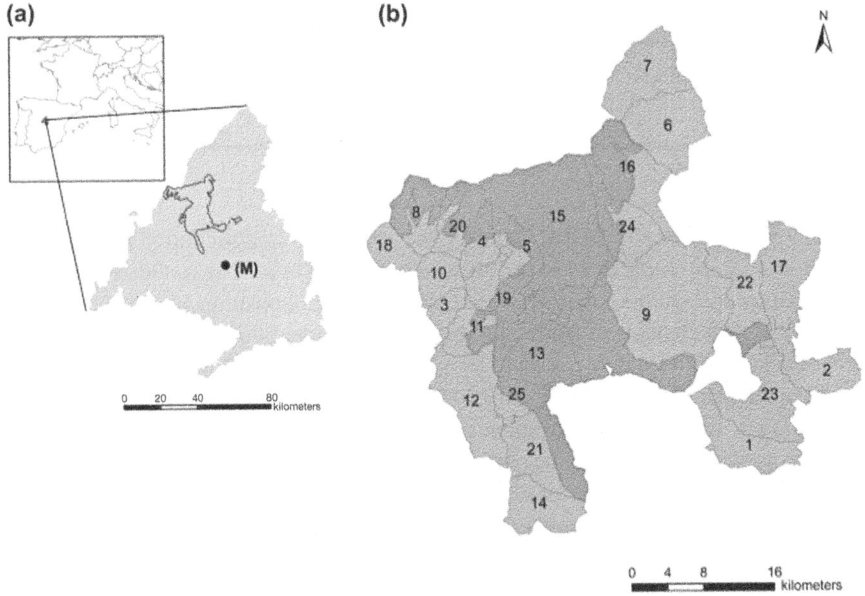

Fig. 6.28 Location of the Cuenca Alta del Manzanares Regional Park, Spain. On the left (a), perimeter of the regional park; M indicates the metropolitan area of Madrid. On the right (b), Cuenca Alta del Manzanares Regional Park (dark grey) and its Socio-economic Influence Zone (light grey). (*Credit*: Schmitz et al. (2012). https://doi.org/10.1016/j.biocon.2012.01.043)

management zones. The Nature Reserve zones occupy 40% of the territory. In them, the permitted uses are traditional forestry and recreational and environmental education activities. Another 47% of the Regional Park is devoted to agricultural and livestock activities. The rest are transition zones – protection areas for valuable and fragile sites – and peri-urban areas surrounding urban centres where some urban development is allowed.

6.3.3.3.2 Protected Biodiversity and Threats

In CAMRP, 13 species of amphibians, 22 species of reptiles, 195 species of birds and 43 species of mammals have been observed. A total of 174 species are included in the National List of Protected Species, 13 of them as Vulnerable (VU) and 7 as Endangered (EN). Endangered species include the black vulture (*Aegypius monachus*), imperial eagle (*Aquila adalberti*), Bonelli's eagle (*Hieraaetus fasciatus*) and lesser kestrel (*Falco naumanni*). Vulnerable species include the eagle owl (*Bubo bubo*) and the golden eagle (*Aquila chrysaetos*). The griffon vulture (*Gyps fulvus*) is a singular and emblematic species of nature conservation in the study area.

Among the unique botanical enclaves, the peaks of the Sierra de Guadarrama are rich in endemic flora. In addition, there are stands of centenary Scots Pine (*Pinus sylvestris*), enclaves of cork oak (*Quercus suber*), a small juniper grove (*Juniperus thurifera*) and some patches of rare or threatened flora. Because of their multifunctionality, traditional agricultural and livestock activities have favoured and preserved this rich and valuable faunal and floristic biodiversity. They have also given rise to a diversity of landscapes: from alpine grasslands and mountain scrublands to mixed agroforestry landscapes of pasture, savannah type and Mediterranean grasslands of the foothills and plains. Thus, one of the objectives of the CAMRP is to conserve rural cultural landscapes.

The most worrying threats for managers and conservationists are population growth, urban sprawl, the construction of new linear infrastructures (motorways and railways), landfills and outdoor recreation. The proximity to the city of Madrid is a factor that hinders the conservation of its rich natural and cultural heritage (Fig. 6.29). On the other hand, Schmitz et al. (2012) emphasise the activation of the natural succession process of vegetation, caused by the abandonment of traditional agricultural activities, especially grazing. Scrublands have invaded former Mediterranean grasslands and part of the dehesas. Despite the large investments of the regional government in forest fire prevention and extinction, the above-mentioned processes have increased the risk of wildfire occurrence and spread (Chuvieco et al. 2010).

Fig. 6.29 Landscape of the Cuenca Alta del Manzanares Regional Park. View from Hoyo de Manzanares to the south. Near the horizon, Monte de El Pardo and the city of Madrid. The abandoned Mediterranean grasslands are being occupied mainly by *Cistus ladanifer* shrubs. (*Credit*: Javier Martínez-Vega)

6.3.3.3.3 Regional Park Effectiveness Assessment

To evaluate the effectiveness of this Regional Park, Schmitz et al. (2012) used a BACI-type method within a complete semi-experimental (CSE) research design. They compared the SIZ (control) with CAMRP (impact). They also compared LULCs 'Before' and After' the park's designation. For this, they took into account a broad time series 10 years before (1975) and 20 years after (2005) the establishment of the Regional Park in 1985. They collected quantitative information on 21 LULCs classes, 22 landscape variables and 19 socio-economic variables and processed them in a GIS. They calculated a land use diversity index, a landscape heterogeneity index and a spatial complexity index that integrates the two former indices. They used a stepwise multiple linear regression model and multivariate analysis to examine changes in rural cultural landscapes (dependent variable) and socio-economic changes (independent variables) of local populations after the establishment of this Regional Park.

6.3.3.3.4 Results

The results show that the assignment of protection to this Regional Park has not prevented the abandonment of traditional agricultural and livestock activities. As a consequence, there has been a process of bush encroachment which has led to a densification and spatial continuity of the forest cover. There has been a process of homogenisation and simplification of the landscape, and the diversity of land uses has decreased. In short, some traditional rural landscapes have been lost. Mediterranean grasslands and dehesas have been the land uses most affected by the processes of change. Considering landscape dynamics, the three indices analysed have worsened after the declaration of the study area as a Regional Park (land use diversity index = −0.491, landscape heterogeneity index = −0.071; spatial complexity index = −0.241). Moreover, they have worsened more than in the surrounding areas (SIZ zone), which include unprotected zones and some zones protected by other PAs, although their protection is more recent (1999 and 2013).

On the other hand, socio-economic patterns have changed from a rural economy to a new non-rural economic model. This is characterised by an increase in population density, the massive use of private vehicles and the development of new communication infrastructures. The latter have led to increased fragmentation of habitats of interest. These new socio-economic patterns affect both the municipalities included in the PA and the surrounding areas. In conclusion, the authors point out that protection policies focused on naturalness and nature marginalised the socio-economic component. Therefore, they recommended incorporating socio-economic components together with ecological indicators in planning and management systems to improve the effectiveness of PAs.

6.3.3.4 Case Study 8: Korup National Park, Cameroon

6.3.3.4.1 Description of the Site

The Korup Forest Reserve (KFR) was designated in 1962, shortly after the independence of Cameroon. The State reserved the right to regulate uses and to allocate 'special privileges' for forestry activities and hunting to local communities. It was later designated as Korup National Park by Presidential Decree No. 86/1283 of 30 October 1986. After its creation, its boundaries were extended to include three more settlements (Nchanji et al. 2021). The aim of the creation of the park was to maintain the rich and unique biodiversity of the area.

Korup National Park (KNP) is located near Cameroon's south-western border with Nigeria (Fig. 6.30). It covers an area of 1,260 km^2. It is one of Cameroon's ten national parks. It connects with the Cross River National Park in Nigeria, another neighbouring national park that protects the extensive rainforest west of the border. The park ranges in altitude from 50 metres to 1,079 metres at Mount Yuhan, the highest peak in the park. The hilly forest landscape is traversed by numerous streams

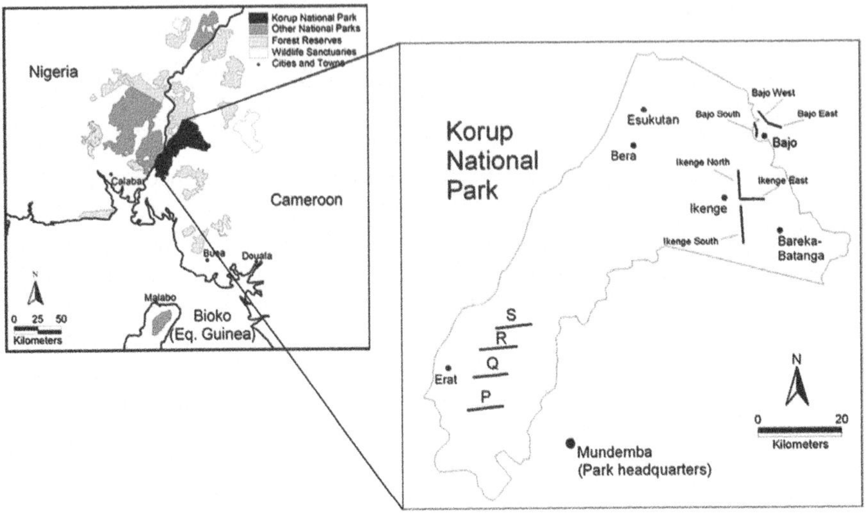

Fig. 6.30 Location of the Korup National Park, Cameroon. In the right inset, the three study sites are observed. They contain the transects along which primate abundance, hunting pressure and habitat characteristics were assessed. (*Credit*: Professor Linder)

that flow into the Korup-Akpasang, Ndian and Bake-Munaya rivers, interspersed with a variety of river habitats.

This national park is representative of the rainforest ecosystem (Oates et al. 2004). It is considered one of the oldest and richest rainforests in Africa in terms of flora and fauna diversity (Fig. 6.31). It is home to more than 1100 species of trees and other vegetation in lower strata. Its canopy reaches 50 metres in height. In the north-eastern part of the park, between Bajo and Ikenge, secondary forest formations predominate. Linder and Oates (2011) studied the impact of hunting on the species of monkeys in this national park (Fig. 6.30) and examined how vulnerability to hunting varies among the different species of monkeys.

6.3.3.4.2 Protected Biodiversity and Threats

The park is home to 161 species of mammals, 430 species of birds and 170 species of amphibians and reptiles (GANP 2021). There are 14 species of primates. Eight of the primates are diurnal and the remaining six are nocturnal. In this study, the authors have focused their attention on seven primate species: putty-nosed monkey (*Cercopithecus nictitans*), crowned monkey (*Cercopithecus pogonias*), red-eared monkey (*Cercopithecus erythrotis*), mona monkey (*Cercopithecus mona*), Preuss's red colobus (*Procolobus pennantii preussi*), red-capped mangabey (*Cercocebus torquatus*) and drill (*Mandrillus leucophaeus*). The red-eared monkey and drill are species endemic to the Cross-Sanaga-Bioko forests, and the subspecies of red

Fig. 6.31 View of the tropical rainforest and the Mana River in Korup National Park, Cameroon. (*Source*: https://national-parks.org/cameroon/korup#&gid=null&pid=3)

colobus are endemic to this part of mainland Africa (Grubb et al. 2003). The red colobus is rated as Critically Endangered (CR), and drill and red-capped mangabey are classified as Endangered (EN) on the IUCN Red List (IUCN 2021b). The red-eared monkey is rated as Vulnerable (VU). The main reported threats are excessive and illegal hunting of wild animals and alteration of the structure and composition of the forest.

6.3.3.4.3 National Park Effectiveness Assessment

To assess the effectiveness of the national park, Linder and Oates (2011) used an ACI-type (After Control-Impact) incomplete semi-experimental research design. Between 2004 and 2005, they compared the primate abundance, hunting pressure and habitat characteristics at three sites in and around the national park. At Mudemba (south of KNP), they surveyed four line transects, labelled P, Q, R and S (Fig. 8.16). At the Ikenge site (in the north-east of KNP), they surveyed three transects. They also surveyed three transects at the Bajo site, outside of KNP. Although the study is one-off (2004–2005), the results were compared with previous studies developed 15 years earlier by other authors (1988/1990; e.g. Infield 1988).

When they found a monkey, they recorded its location, species and the number of individuals sighted. These data were transformed into a measure of relative density, the frequency of sightings of each species. They also assessed the habitat structure and floristic composition of the trees distributed along each of the above-mentioned transects. They collected data on their location, diameter at breast height (dbh), family and species. To quantify spatial variation in hunting intensity, they recorded evidence of hunting (traps, cartridges, hunters' camps, etc.) in two subperiods (May–June 2004 and October 2004–June 2005) along the same transects. They also conducted hunter surveys that they compared with those conducted 15 years earlier. Their aim was to find out how the proportional representation of primate species in hunters' harvests had changed over time.

To assess the vulnerability of species to hunting and the impact of hunting on monkey community structure, they tested three methods and corresponding indicators. The first compared primate abundance between areas that varied in hunting pressure but were similar in floristic structure and composition. The second method quantified temporal changes (1992–2005) in primate sighting frequencies in Ikenge. They used a third method to examine temporal changes in the proportional representation of different primate species hunted in Ikenge case.

6.3.3.4.4 Results

According to the hunting response index, only the putty-nosed and mona monkeys are relatively tolerant to heavy hunting. The remaining five species showed a high vulnerability to hunting, especially drill and mangabey. Considering the temporal dynamics (1990–2004/2005) in primate abundance at the Ikenge site, the authors noted a decline in primate species richness. The percentage of sightings of red colobus, crowned monkeys and drills (which were not sighted at Ikenge in 2004/2005) decreased over time. Finally, the authors indicated that the species most captured in Ikenge were the putty-nosed monkey, Preuss's red colobus and mona monkey. Over time, they recorded a significant decrease of red colobus and drills in relation to total captures. On the contrary, the proportional representation of *Cercopithecus* species increased over the period analysed.

In summary, according to Linder and Oates (2011), hunting is causing worrying alterations to Africa's most threatened primate species. For example, Preuss's red colobus and the drill are on their way to local extinction in the KNP. Among the ineffectiveness factors, they highlight the following: lack of rigorous surveillance and control of primate hunting, especially of the most endangered species; absence of ambitious environmental awareness programmes; or lack of a well-implemented socio-economic support programme for the local population. Linder and Oates (2011) made the following recommendations to KNP managers, accordingly: focusing NP management efforts in areas far from human settlements and where hunting levels are already quite low; streamlining NP management efforts in areas where the richness and abundance of primate species is the greatest; initiating conservation awareness and education programmes in villages located within and surrounding to NP; and implementing a long-term and consistent biological monitoring programme.

6.3.4 Lessons Learned from Failure Cases

Here we summarise the commonly found factors for the lack of effectiveness of the selected cases and the recommendations that the authors made in order to overcome existing deficiencies.

6.3.4.1 Ineffectiveness Factors

Some common characteristics among the cases whose results are negative or tend to worsen arose:

1. *Legal stringency*. In the cases of failure, we have not found any differentiating pattern depending on the type of protection. In the selected literature, there are cases of ineffectiveness both in protected areas of maximal legal stringency (reserves, Russ et al. 2017; Zang et al. 2017; and national parks, Linder and Oates 2011) and in multiple-use (M)PAs (Schmitz et al. 2012).
2. *Convergence of multiple pressures in and around (M)PAs*. Although some authors (Russ et al. 2017) recognized that natural phenomena also cause disturbances to PAs (e.g. typhoons in marine reserves), most of these pressures were anthropogenic. Illegal fishing (Russ et al. 2017); high pressure from hunting (Linder and Oates 2011) and fishing (Russ et al. 2017); expansion of farmland and aquaculture water areas and invasion of alien species (Zang et al. 2017); urban expansion and infrastructure construction (Schmitz et al. 2012), among others, are some socio-economic activities that alter the ecological balance of PAs. Decrease in species abundance and diversity (Russ et al. 2017), degradation of wetlands and water quality (Zang et al. 2017) and fragmentation of landscapes (Schmitz et al. 2012) are some of the effects caused by the above-mentioned pressures. On the contrary, the abandonment of traditional agricultural and livestock activities in marginal rural areas implies other simultaneous impacts such as the encroachment of scrub on livestock areas (Schmitz et al. 2012), the homogeneity of the landscape and the increase of fuel loads and their continuity.

6.3.4.2 Recommendations

1. *Consistent, continuous, and long-term monitoring and tracking programs* for target habitats and species are considered essential (Linder and Oates 2011; Russ et al. 2017), both within protected areas and in their immediate surroundings.
2. *Focus and streamline management efforts* on key habitats and areas where target species richness and abundance are greatest (Linder and Oates, 2011).
3. Improve the effectiveness of PAs by considering an *adaptive management system* that harmonizes biodiversity conservation and socio-economic development (Zang et al. 2017).

4. *Incorporate socioeconomic components* along with ecological indicators *into planning and management systems* to improve PA effectiveness (Schmitz et al. 2017).
5. *Encourage traditional agroforestry-livestock activities* that have been architects of biodiversity conservation for centuries, especially in multiple-use PAs (Schmitz et al. 2012). On the contrary, it is advisable to regulate and monitor new land uses and infrastructures that fragment and isolate ecosystems of higher ecological value and increase the risks of degradation of natural habitats (Schmitz et al. 2012).
6. *Initiate environmental education and awareness programs* to promote biodiversity conservation in villages and municipalities located in and around PAs (Linder and Oates 2011).
7. Finally, *natural disturbances or risks* (typhoons and others related to climate change) *should be anticipated and managed* as much as possible to minimize their impacts on biodiversity (Russ et al. 2017).

Chapter 7
The Future of Protected Areas and Biodiversity

(M)PAs help to conserve biodiversity. They are, however, just part of the solution to the current environmental crisis that characterises the Anthropocene (Crutzen 2002). If the current trends of biodiversity continue, we are shortly facing a dramatic decline of life on Earth that is already under way. Such decline will do away with plenty of unique, extraordinary forms of life and leave our Planet and ourselves more empty and vulnerable. If the right to exist of every other species was not convincing enough, we should also carefully consider the foreseen large decrease in human well-being that such biological impoverishment will entail (IPBES 2019). By eliminating species, we are eliminating our well-being and chances for survival too.

We are fighting a battle against the clock. Human-made pressures leading to ecosystem destruction and species' extinction are not only taking place at an unprecedented speed but are also on the rise over a constantly shrinking natural environment. The first alarms on the tremendous damage humans were inflicting upon the rest of the natural world rang already 50 years ago (UNDPA 1972). Efforts have since been made to expand the global (M)PA network to cover almost 17% and 8% of terrestrial and marine ecosystems, respectively (Protected Planet 2020). They have, however, been largely insufficient to counter mounting pressures and ensure a healthy ecological balance of the Planet (IPBES 2019; WWF 2020). One cannot help wondering whether, if we had had better political leadership, greater will and more determination on the one hand and less greed, short-sightedness and selfishness on the other hand, we could now be in a much better position, both biodiversity-wise and human-wise. Too many empty statements have been made. Too many useless agreements have been signed. Far too much time has been wasted. We cannot afford to waste a single more minute if we want to amend the ecological mess we have created. To do this, a multi-pronged approach whereby 'passive' conservation tools such as (M)PAs or gene banks are strongly complemented by 'active', large-scale ecosystem restoration will most likely be needed to retain as much biodiversity as possible before it is gone forever.

D. Rodríguez-Rodríguez, J. Martínez-Vega, *Effectiveness of Protected Areas in Conserving Biodiversity*, Strategies for Sustainability,
https://doi.org/10.1007/978-3-030-94297-7_7

Fig. 7.1 Wolf skins hanging from a yurt. Gorkhi-Terelj National Park, Mongolia. (*Credit*: David Rodríguez)

We have to rapidly expand the global network of (M)PA and OECMs to ecologically meaningful and environmentally sustainable levels. It does not seem too wild or generous, but probably just fair and sensible, to reserve at least half of the Planet for the other species to thrive and evolve, whereas sustainably managing the other half for human well-being, as suggested previously (Wilson 2016; Dudley et al. 2018). In this sense, the foreseeable new international biodiversity protection coverage targets at 30% (CBD 2021), though praiseworthy, will likely be insufficient both in terms of area and stringency, according to current evidence (IPBES 2019). Unlike much previous (M)PA coverage expansion, such enlargement of protection should target biodiversity hotspots and Key Biodiversity Areas (Hoffman et al. 2016) where rich, threatened and endemic biodiversity dwell.

However, just legally designating new sites, even if under restrictive legal schemes, is unlikely to fully achieve conservation targets, especially in pressured large, remote areas subject to weak institutional capacity or support. Thus, targeted protection expansion must be coupled with increased means and capacities for managing existing and new protected sites (Minin and Toivonen 2015). Two managerial activities make the backbone of (M)PA effectiveness and must therefore be ensured in all protected sites: monitoring and surveillance. They are essential to spot and stop inner pressures as well as to identify conservation successes and act upon failures (Bergseth et al. 2018; Miller et al. 2013).

Fig. 7.2 Critically endangered indri in Andasibe-Mantadia National Park, Madagascar. (*Credit*: David Rodríguez)

Increased efficiency of existing funds and additional funding will be required for this. Scarce resources for (M)PA management should be put to their best use by: focusing research and conservation actions, including monitoring, on key biodiversity (i.e. threatened or endemic genes, species and habitats), rather than on easily-tracked or fashionable biodiversity; promoting ongoing practices leading to effective biodiversity conservation and use through OECM recognition; upholding community-based management and engagement; implementing scientifically-driven monitoring protocols; seeking collaboration with national and international research and education institutions; and assessing and, where needed, strengthening, relocating or degazzeting ineffective (M)PAs.

One way to increase ever-lacking funds for (M)PAs is developing business plans for (M)PAs including publicity, branding, (eco)labelling and marketing, to make them largely economically self-sufficient (Dharmaratne et al. 2000). Another complementary way is seeking public-private partnerships for funding, including land stewardship schemes and sponsoring (de la Varga Pastor and Pons Solé 2018). A third way, already in use in many countries, is introducing entrance fees (Atmodjo et al. 2017) or other fees for the services provided by the (M)PAs such as guides, visitors' centres or parking lots (Barcellini et al. 2013). International donors and agencies' support could also be sought (Lindsey et al. 2021). Last but not least, governments should increase their financial commitments to preserving life on Earth by financially ensuring at least basic managerial operations in all (M)PAs.

Some studies predict that climate change will put additional pressure on threatened species over the next 50 years (Ebrahimi et al. 2021). Whereas increased coverage of (M)PAs is believed to significantly increase the resilience of biological communities to climate change (Lehikoinen et al. 2021), common principles of climate change adaptation need to be adequately incorporated into the management plans of many (M)PAs around the world, especially of heavily affected ecosystems such as mountains or wetlands (O'Regan et al. 2021). Given that many species are expected to respond to climate change by shifting their distribution ranges to climatically suitable areas, conservation planning needs to be approached in a dynamic, adaptive manner which may require relocation of some existing (M)PAs (Coldrey and Turpei 2021).

As mentioned, (M)PAs are intended to conserve biodiversity on site. In the settings where that is no longer possible (which exposes a severe failure as society), emergency tools such as seed banks, gene banks, botanical gardens or zoos will remain the last safety net for many species unrightfully expelled from their home ranges. Whereas protected sites and *ex situ* facilities will remain essential, complementary tools to reduce biodiversity loss, they remain largely 'defensive' strategies bound to produce suboptimal outcomes from a conservation viewpoint (Butchart et al. 2010).

If we aim to truly reverse ecological degradation, we need to make more quality spaces available for other species, both by increasing protection of valuable remaining habitats and by 'reclaiming' and conditioning new spaces of little use to wildlife currently as new protected sites or ecological corridors (Mingarro and Lobo 2021). This is the combined conservation approach envisaged for the UN's Decade on Ecological Restoration in its Vision/Agenda for 2030 that heavily relies on ecosystem restoration to reverse current negative biodiversity trends (UN 2021). This approach has been taken up by international institutions such as the European Union, which has committed to quantitative restoration targets by 2030 (European Commission 2020b). We must take into account that restored ecosystems seldom have the same ecological characteristics as original ecosystems. Moreover, ecological restoration can be very expensive. Thus, from a conservation and economic standpoint, conserving natural habitats will always be preferable to restoring degraded biodiversity.

We have the knowledge, technology and means to start acting differently, both as society and as individuals. We know that major pressures to biodiversity and also biodiversity protection largely result from political decisions (not acting is a decision too). It is high time that societies around the World united to defend their right

Fig. 7.3 Captivity breeding is essential to maintain the World's biodiversity. (*Credit*: David Rodríguez)

to live in a healthy and ecologically productive Planet and claimed other species' right to exist before short-sighted decision-makers and greedy businessmen.

Nevertheless, biodiversity conservation has also much to do with individual behaviours. Most of us in developed societies are consuming too much. Far more than the Planet can provide, and we are doing so at the expense of other species. We have to learn to live reasonably well so that other species can simply live. That entails giving up excess. Overdeveloping in a finite Planet is not an option. In the end, biodiversity conservation will depend on everyone of us. We must finally come to terms with the reality of a finite Planet that cannot stand evergrowing human population and endless consumption. We must adapt to living within the ecological boundaries that have sustained life on Earth for millions of years, and we must do it quickly.

Ecologically, humans are the undisputed dominant species. We dwell in almost every habitat, and predate, manage or outcompete every other living species. However superior and detached from nature we think we are, the reality is that we depend on biodiversity. We eat biodiversity; we breathe biodiversity; we dress biodiversity; we build with biodiversity; we visit biodiversity in our free time; we enjoy biodiversity in the wild and at home; we are inspired by biodiversity. Biodiversity is essential to human well-being and to our own survival as species, because we are biodiversity.

Fig. 7.4 The future of biodiversity is in our hands. (*Credit*: David Rodríguez)

Appendix 1

Articles Included in the Review[1]

Rodríguez-Rodríguez D, Sebastiao J, Salvo Tierra AE et al. (2019) Effect of protected areas in reducing land development across geographic and climate conditions of a rapidly developing country, Spain. Land Degradation & Development 30: 991–1005. https://doi.org/10.1002/ldr.3286

Hellwig N, Walz A, Markovic D (2019) Climatic and socioeconomic effects on land cover changes across Europe: Does protected area designation matter? PLoS ONE 14(7): e0219374. https://doi.org/10.1371/journal.pone.0219374

Guidetti P, Addis P, Atzori F et al. (2019) Assessing the potential of marine Natura 2000 sites to produce ecosystem-wide effects in rocky reefs: A case study from Sardinia Island (Italy). Aquatic Conservation: Marine & Freshwater Environment 29: 537–545. https://doi.org/10.1002/aqc.3026

da Silva C, Zenóbia de Oliveira L, Balée W et al. (2019) Human impact on the abundance of useful species in a protected area of the Brazilian Cerrado by people perception and biological data. Landscape Research 44(1): 75–87. https://doi.org/10.1080/01426397.2017.1396304

Anderson A, Batista M, Gibran F et al. (2019) Habitat use of five key species of reef fish in rocky reef systems of southern Brazil: evidences of MPA effectiveness. Marine Biodiversity 49: 1027–1036. https://doi.org/10.1007/s12526-018-0893-6

Kilungu H, Leemans R, Munishi P et al. (2019) Forty Years of Climate and Land-Cover Change and its Effects on Tourism Resources in Kilimanjaro National Park. Tourism Planning & Development 16 (2): 235–253. https://doi.org/10.1080/21568316.2019.1569121

[1] In chronological order of appearance (newest first).

D. Rodríguez-Rodríguez, J. Martínez-Vega, *Effectiveness of Protected Areas in Conserving Biodiversity*, Strategies for Sustainability,
https://doi.org/10.1007/978-3-030-94297-7

Eddy TD, Friedlander AM, Salinas de León P (2019). Ecosystem effects of fishing & El Niño at the Galápagos Marine Reserve. PeerJ 7:e6878. https://doi.org/10.7717/peerj.6878

Llopis J, Harimalala P, Bär R et al. (2019) Effects of protected area establishment and cash crop price dynamics on land use transitions 1990–2017 in north-eastern Madagascar. Journal of Land Use Science 14: 52–80. https://doi.org/10.1080/1747423X.2019.1625979

Merkohasanaj M, Rodríguez-Rodríguez D, García Martínez M et al. (2019) Assessing the environmental effectiveness of the Spanish Marine Reserve Network using remote sensing. Ecological Indicators 107: 105583. https://doi.org/10.1016/j.ecolind.2019.105583

Yuh YG, Dongmo ZN, N'Goran PK et al. (2019) Effects of Land cover change on Great Apes distribution at the Lobéké National Park and its surrounding Forest Management Units, South-East Cameroon. A 13 year time series analysis. Sci Rep 9: 1445. https://doi.org/10.1038/s41598-018-36225-2

Qian D, Cao G, Du Y et al. (2019) Impacts of climate change and human factors on land cover change in inland mountain protected areas: a case study of the Qilian Mountain National Nature Reserve in China. Environmental Monitoring and Assessment 191(8): 486. https://doi.org/10.1007/s10661-019-7619-5

Zhu P, Cao W, Huang L et al. (2019) The Impacts of Human Activities on Ecosystems within China's Nature Reserves. Sustainability 11(23): 6629. https://doi.org/10.3390/su11236629

Mona M, El-Naggar H, El-Gayar E et al. (2019) Effect of human activities on biodiversity in Nabq Protected Area, South Sinai, Egypt. The Egyptian Journal of Aquatic Research 45(1): 33–43. https://doi.org/10.1016/j.ejar.2018.12.001

Biró M, Bölöni J, Molnár Z (2018) Use of long-term data to evaluate loss and endangerment status of Natura 2000 habitats and effects of protected areas. Conservation Biology 32(3): 660–671. https://onlinelibrary.wiley.com/doi/full/10.1111/cobi.13038

Rodríguez-Rodríguez D, Martínez-Vega J (2018a) Protected area effectiveness against land development in Spain. Journal of Environmental Management 215: 345–357. https://doi.org/10.1016/j.jenvman.2018.03.011

Salghuna N, Rama Chandra Prasad R, Asha Kumari J (2018) Assessing the impact of land use and land cover changes on the remnant patches of Kondapalli reserve forest of the Eastern Ghats, Andhra Pradesh, India. The Egyptian Journal of Remote Sensing and Space Science 24(2): 329–330. https://doi.org/10.1016/j.ejrs.2018.01.005

Muñoz Brenes CL, Jones KW, Schlesinger P et al. (2018) The impact of protected area governance and management capacity on ecosystem function in Central America. PLoS One 13(10): e0205964. https://doi.org/10.1371/journal.pone.0205964

Jin Y, Fan H (2018) Land use/land cover change and its impacts on protected areas in Mengla County, Xishuangbanna, Southwest China. Environmental Monitoring and Assessment 190(9): 509. https://doi.org/10.1007/s10661-018-6891-0

Beresford A, Buchanan G, Phalan B et al. (2018) Correlates of long-term land-cover change and protected area performance at priority conservation sites in Africa. Environmental Conservation 45(1): 49–57. https://doi.org/10.1017/S0376892917000157

Egerton JP, Johnson AF, Turner J et al. (2018) Hydroacoustics as a tool to examine the effects of Marine Protected Areas and habitat type on marine fish communities. Science Reports 8:47. https://doi.org/10.1038/s41598-017-18353-3

da Silva M, Paviolo A, Reverberi Tambosi L et al. (2018) Effectiveness of Protected Areas for biodiversity conservation: Mammal occupancy patterns in the Iguaçu National Park, Brazil. Journal for Nature Conservation 41: 51–62. https://doi.org/10.1016/j.jnc.2017.11.001

Starrs C, Butsic V, Stephens C et al. (2018) The impact of land ownership, firefighting, and reserve status on fire probability in California. Environmental Research Letters 13: 034025. https://doi.org/10.1088/1748-9326/aaaad1

Rodríguez-Rodríguez D, Martínez-Vega J (2018b) Effect of legal protection and management of protected areas at preventing land development: a Spanish case study. Reg Environ Change 18: 2483–2494. https://doi.org/10.1007/s10113-018-1369-8

Ngoc QTK (2018) Impacts on the ecosystem and human well-being of the marine protected area in Cu Lao Cham, Vietnam. Marine Policy 90: 174–183. https://doi.org/10.1016/j.marpol.2017.12.015

Tapia-Armijos MF, Homeier J, Munt DD (2017) Spatio-temporal analysis of the human footprint in South Ecuador: Influence of human pressure on ecosystems and effectiveness of protected areas. Applied Geography 78: 22–32. https://doi.org/10.1016/j.apgeog.2016.10.007

Emets VM (2017) On assessment of the conservation effectiveness of Red Data insect species in the territories of a nature reserve and a nature sanctuary within the same reservation area (by the example of Voronezh Biosphere Nature Reservation). Entomological Review 97: 20–29. https://doi.org/10.1134/S0013873817010043

McKenna M, Gabriele C, Kipple B (2017) Effects of marine vessel management on the underwater acoustic environment of Glacier Bay National Park, AK. Ocean & Coastal Management 139: 102–112. https://doi.org/10.1016/j.ocecoaman.2017.01.015

McCune JL, Van Natto A, MacDougall AS (2017) The efficacy of protected areas and private land for plant conservation in a fragmented landscape. Landscape Ecology 32: 871–882. https://doi.org/10.1007/s10980-017-0491-1

Bonelli M, Manenti R, Scaccini D (2017) Mountain protected areas as refuges for threatened freshwater species: the detrimental effect of the direct introduction of alien species. Journal on Protected Mountain Areas Research and Management 1: 23–29. https://doi.org/10.1553/eco.mont-9-2s23

Russ GR, Lowe JR, Rizzari JR et al. (2017) Partitioning no-take marine reserve (NTMR) and benthic habitat effects on density of small and large-bodied tropical wrasses. PLoS One 12 (12): e0188515. https://doi.org/10.1371/journal.pone.0188515

Zang Z, Zou X, Zuo P et al. (2017) Impact of landscape patterns on ecological vulnerability and ecosystem service values: An empirical analysis of Yancheng Nature Reserve in China. Ecological Indicators 72:142–152. https://doi.org/10.1016/j.ecolind.2016.08.019

Friedlander AM, Golbuu Y, Ballesteros E et al. (2017) Size, age, and habitat determine effectiveness of Palau's Marine Protected Areas. PLoS One. 12(3): e0174787. https://doi.org/10.1371/journal.pone.0174787

Vergara P, Meneses LO, Saavedra M et al. (2017) Magellanic Woodpeckers in three national parks of central-southern Chile: habitat effects and population variation over the last two decades. Avian Conservation and Ecology 12(2): 15. https://doi.org/10.5751/ACE-01106-120215

Gong M, Fan Z, Zhang X et al. (2017) Measuring the effectiveness of protected area management by comparing habitat utilization and threat dynamics. Biological Conservation 210: 253–260. https://doi.org/10.1016/j.biocon.2017.04.027

Trowbridge CD, Kachmarik K, Plowman CQ et al. (2017) Biodiversity of shallow subtidal, under-rock invertebrates in Europe's first marine reserve: Effects of physical factors and scientific sampling. Estuarine, Coastal and Shelf Science 187: 43–52. https://doi.org/10.1016/j.ecss.2016.12.022

Hiley JR, Bradbury RB, Thomas CD (2016) Impacts of Habitat Change and Protected Areas on Alpha and Beta Diversity of Mexican Birds. Diversity & Distributions 22(12): 1245–1254. https://doi.org/10.1111/ddi.12483

Watson HV, McKeown NJ, Coscia I et al. (2016) Population genetic structure of the European lobster (Homarus gammarus) in the Irish Sea and implications for the effectiveness of the first British marine protected area. Fisheries Research 183: 287–293. https://hdl.handle.net/10.1016/j.fishres.2016.06.015

Lagesse J, Thondhlana G (2016) The effect of land-use on small mammal diversity inside and outside the Great Fish River Nature Reserve, Eastern Cape, South Africa. Journal of Arid Environments 130: 76–83. https://doi.org/10.1016/j.jaridenv.2016.03.006

Sweke EA, Assam JM, Chande A et al. (2016) Comparing the Performance of Protected and Unprotected Areas in Conserving Freshwater Fish Abundance and Biodiversity in Lake Tanganyika, Tanzania. International Journal of Ecology 2016: 7139689. https://doi.org/10.1155/2016/7139689

Schmidt M, Mbayngone E, Bachmann Y et al. (2016) The impact of land use on species composition and habitat structure in Sudanian savannas — A modelling study in protected areas and agricultural lands of southeastern Burkina Faso. Candollea 71: 265–274. https://doi.org/10.15553/c2016v712a11

Kroner RE, Krithivasan R, Mascia M (2016) Effects of protected area downsizing on habitat fragmentation in Yosemite National Park (USA), 1864–2014. Ecology and Society 21(3): 22. https://doi.org/10.5751/ES-08679-210322

Howarth LM, Roberts CM, Hawkins JP et al. (2015) Effects of ecosystem protection on scallop populations within a community-led temperate marine reserve. Marine Biology 162: 823–840. https://doi.org/10.1007/s00227-015-2627-7

Hermoso V, Filipe AF, Segurado P et al. (2015) Effectiveness of a large reserve network in protecting freshwater biodiversity: a test for the Iberian Peninsula. Freshwater Biology 60(4): 698–710. https://doi.org/10.1111/fwb.12519

Russ GR, Miller K, Rizzari JR et al. (2015a) Long-term no-take marine reserve and benthic habitat effects on coral reef fishes. Marine Ecology Progress Series 529: 233–248. https://www.jstor.org/stable/24895912

Harmelin-Vivien M, Cottalorda JM, Dominici JM et al. (2015) Effects of reserve protection level on the vulnerable fish species Sciaena umbra and implications for fishing management and policy. Global Ecology and Conservation 3: 279–287. https://doi.org/10.1016/j.gecco.2014.12.005

Françoso RD, Brandão RA, Nogueira C et al. (2015) Habitat loss and the effectiveness of protected areas in the Cerrado Biodiversity Hotspot. Natureza & Conservacao 13: 35–40. https://doi.org/10.1016/j.ncon.2015.04.001

Russ GR, Bergseth BJ, Rizzari JR et al (2015b) Decadal-scale effects of benthic habitat and marine reserve protection on Philippine goatfish (F: Mullidae). Coral Reefs 34: 773–787. https://doi.org/10.1007/s00338-015-1296-9

Huang TCC, Lo KFA (2015) Effects of Land Use Change on Sediment and Water Yields in Yang Ming Shan National Park, Taiwan. Environments 2(1): 32–42. https://doi.org/10.3390/environments2010032

Remteng C, Adepoju KA, Akinyede JO (2015) Geospatial Analysis of Fragmentation and its Effects on Biodiversity: A Case Study of Reserve Forest. International Journal of Ecological Economics & Statistics 36(1). http://ceser.in/ceserp/index.php/ijees/article/view/3396

Na XD, Zang SY, Zhang NN et al. (2015) Impact of land use and land cover dynamics on Zhalong wetland reserve ecosystem, Heilongjiang Province, China. International Journal of Environmental Science and Technology 12: 445–454. https://doi.org/10.1007/s13762-013-0398-6

Velho N, Agarwala M, Srinivasan U et al. (2014) Collateral damage: impacts of ethno-civil strife on biodiversity and natural resource use near Indian nature reserves. Biodiversity Conservation 23: 2515–2527. https://doi.org/10.1007/s10531-014-0735-1

Carranza T, Balmford A, Kapos V et al. (2014) Protected Area Effectiveness in Reducing Conversion in a Rapidly Vanishing Ecosystem: The Brazilian Cerrado. Conservation Letters 7(3): 216–223. https://doi.org/10.1111/conl.12049

Terra TN, dos Santos RF, Costa DV (2014) Land use changes in protected areas and their future: The legal effectiveness of landscape protection. Land Use Policy 38: 378–387. https://doi.org/10.1016/j.landusepol.2013.12.003

Satumanatpan S, Senawongse P, Thansuporn W et al. (2014) Enhancing management effectiveness of environmental protected areas, Thailand. Ocean & Coastal Management 89: 1–10. https://doi.org/10.1016/j.ocecoaman.2013.12.001

Byers A (2014) Contemporary Human Impacts on Subalpine and Alpine Ecosystems of the Hinku Valley, Makalu-Barun National Park and Buffer Zone, Nepal. Himalaya. The Journal of the Association for Nepal and Himalayan Studies 33(1): 8. http://digitalcommons.macalester.edu/himalaya/vol33/iss1/8

Machumu ME, Yakupitiyage A (2013) Effectiveness of marine protected areas in managing the drivers of ecosystem change: a case of Mnazi Bay Marine Park, Tanzania. Ambio 42(3): 369–80. https://doi.org/10.1007/s13280-012-0352-8

Sieber A, Kuemmerle T, Prishchepov AV et al. (2013) Landsat-Based Mapping of Post-Soviet Land-Use Change to Assess the Effectiveness of the Oksky and Mordovsky Protected Areas in European Russia. Remote Sensing of Environment 133: 38–51. https://doi.org/10.1016/j.rse.2013.01.021

Green JMH, Larrosa C, Burgess ND et al. (2013) Deforestation in an African biodiversity hotspot: Extent, variation and the effectiveness of protected areas. Biological Conservation, 164: 62–72. https://doi.org/10.1016/j.biocon.2013.04.016

Rodríguez N, Armenteras D, Retana J (2013) Effectiveness of protected areas in the Colombian Andes: Deforestation, fire and land-use changes. Regional Environmental Change, 13(2): 423–435. https://doi.org/10.1007/s10113-012-0356-8

Petrosillo I, Semeraro T, Zaccarelli N et al. (2013) The possible combined effects of land-use changes and climate conditions on the spatial-temporal patterns of primary production in a natural protected area. Ecological Indicators 29: 367–375. https://doi.org/10.1016/j.ecolind.2013.01.025

Alexander TJ, Gladstone W (2013) Assessing the effectiveness of a long-standing rocky intertidal protected area and its contribution to the regional conservation of species, habitats and assemblages. Aquatic Conservation-marine and Freshwater Ecosystems 23: 111–123. https://doi.org/10.1002/aqc.2284

Andam KS, Ferraro PJ, Hanauer MM (2013) The effects of protected area systems on ecosystem restoration: a quasi-experimental design to estimate the impact of Costa Rica's protected area system on forest regrowth. Conservation Letters 6(5): 317–323. https://doi.org/10.1111/conl.12004

Ohnesorge B, Plieninger T, Hostert P (2013) Management effectiveness and land cover change in dynamic cultural landscapes—assessing a Central European biosphere reserve. Ecology and Society 18(4): 23. https://doi.org/10.5751/ES-05888-180423

Olds A, Connolly R, Pitt K et al. (2012) Habitat connectivity improves reserve performance. Conservation Letters 5: 56–63. https://doi.org/10.1111/j.1755-263X.2011.00204.x

López C, Poladura A, Hernández JC et al. (2012) Contrasting effects of protection from harvesting in populations of two limpet species in a recently established marine protected area. Scientia Marina 76 (4): 799–807. https://doi.org/10.3989/scimar.03601.15C

Nautiyal S, Nidamanuri RR (2012) Ecological and socioeconomic impacts of conservation policies in biodiversity hotspots: a case study from Rajiv Gandhi National Park, India. Journal of Environmental Studies and Sciences 2: 165–177. https://doi.org/10.1007/s13412-011-0052-x

Gaveau DLA, Curran LM, Paoli GD et al. (2012) Examining protected area effectiveness in Sumatra: Importance of regulations governing unprotected lands.

Conservation Letters 5 (2): 142–148. https://doi.org/10.1111/j.1755-263X.2011.00220.x

Schmitz MF, Matos DGG, De Aranzabal I et al. (2012) Effects of a protected area on land-use dynamics and socioeconomic development of local populations. Biological Conservation 149(1): 122–135. https://doi.org/10.1016/j.biocon.2012.01.043

Xie Z, Liu J, Ma Z et al. (2012) Effect of surrounding land-use change on the wetland landscape pattern of a natural protected area in Tianjin, China. International Journal of Sustainable Development & World Ecology 19(1): 16–24. https://doi.org/10.1080/13504509.2011.583697

Dong H, Li Y, Wang Q et al. (2011) Impacts of invasive plants on ecosystems in natural reserves in Jiangsu of China. Russian Journal of Ecology 42: 133–137. https://doi.org/10.1134/S1067413611020044

Nacoulma BMI, Schumann K, Traoré S et al. (2011) Impacts of land-use on West African savanna vegetation: a comparison between protected and communal area in Burkina Faso. Biodiversity & Conservation 20: 3341–3362. https://doi.org/10.1007/s10531-011-0114-0

Linder J, Oates J (2011) Differential impact of bushmeat hunting on monkey species and implications for primate conservation in Korup National Park, Cameroon. Biological Conservation, 144: 738–745. https://doi.org/10.1016/j.biocon.2010.10.023

Rodríguez-Loinaz G, Amezaga I, Onaindia M (2011) Efficacy of Management Policies on Protection and Recovery of Natural Ecosystems in the Urdaibai Biosphere Reserve. Natural Areas Journal 31(4): 358–367. https://doi.org/10.3375/043.031.0406

Simelane TS (2010) Impacts of traditional land uses on biodiversity outside conservation areas: effects on dung beetle communities of Vaalbos National Park. African Journal of Ecology 48(2): 490–501. https://doi.org/10.1111/j.1365-2028.2009.01138.x

Tkachenko KS, Soong K (2010) Protection of habitat types: a case study of the effectiveness of a small marine reserve and impacts of different habitats on the diversity and abundance of coral reef fishes. Zoological Studies 49: 195–210. http://zoolstud.sinica.edu.tw/Journals/49.2/195.pdf

Jaworski A, Sólmundsson J, Ragnarsson S (2010) Fish assemblages inside and outside marine protected areas of northern Iceland: protection effects or environmental confounds? Fisheries Research 102: 50–59. https://doi.org/10.1016/j.fishres.2009.10.008

References

Adams VM, Iacona GD, Possingham HP (2019) Weighing the benefits of expanding protected areas versus managing existing ones. Nature Sustainability 2: 404–411. https://doi.org/10.1038/s41893-019-0275-5

Addison P (2011) A global review of long-term Marine Protected Area monitoring programmes: The application of a good framework to marine biological monitoring. JNCC, Peterborough. https://data.jncc.gov.uk/data/43cd05f6-ca97-4cdb-9f19-3bb14197adce/JNCC-Report-455-FINAL-WEB.pdf

Addison PFE, Flander LB, Cook CN (2015) Are we missing the boat? Current uses of long-term biological monitoring data in the evaluation and management of marine protected areas. Journal of Environmental Management 149: 148–156. https://doi.org/10.1016/j.jenvman.2014.10.023

Alcala AC, Russ GR (2006) No-take marine reserves and reef fisheries management in the Philippines: a new people power revolution. Ambio 35(5): 245–254. https://doi.org/10.1579/05-a-054r1.1

Almassi B (2017) Ecological Restorations as Practices of Moral Repair. Ethics and the Environment 22: 19–40. https://www.jstor.org/stable/10.2979/ethicsenviro.22.1.02

Andam KS, Ferraro PJ, Pfaff A et al. (2008) Measuring the effectiveness of protected area networks in reducing deforestation. PNAS 105: 16089–16094. https://doi.org/10.1073/pnas.0800437105

Arendt F, Matthes J (2014) Nature Documentaries, Connectedness to Nature, and Pro-environmental Behavior. Environmental Communication 10: 453–472. https://doi.org/10.1080/17524032.2014.993415

Atmodjo E, Lamers M, Mol A (2017) Financing marine conservation tourism: Governing entrance fees in Raja Ampat, Indonesia. Marine Policy 78: 181–188. https://doi.org/10.1016/j.marpol.2017.01.023

Azqueta D, Alviar M, Domínguez L, O'Ryan R (2007) Introducción a la Economía Ambiental. Segunda Edición. McGraw-Hill Interamericana, Madrid, Spain

Balmford A, Gravestock P, Hockley N et al. (2004) The worldwide costs of marine protected areas. PNAS 101(26): 9694–9697. https://doi.org/10.1073/pnas.0403239101

Barcellini VC, Motta FS, Martins AM et al. (2013) Recreational anglers and fishing guides from an estuarine protected area in southeastern Brazil: Socioeconomic characteristics and views on fisheries management. Ocean & Coastal Management 76: 23–29. https://doi.org/10.1016/j.ocecoaman.2013.02.012

Barnes M, Craigie I, Harrison L et al. (2016) Wildlife population trends in protected areas predicted by national socio-economic metrics and body size. Nature Communications 7: 12747. https://doi.org/10.1038/ncomms12747

D. Rodríguez-Rodríguez, J. Martínez-Vega, *Effectiveness of Protected Areas in Conserving Biodiversity*, Strategies for Sustainability,
https://doi.org/10.1007/978-3-030-94297-7

Barnosky AD (2008) Megafauna Biomass Tradeoff as a Driver of Quaternary and Future Extinctions. PNAS 105 (suppl. 1): 11543–11548. http://www.pnas.org/content/pnas/105/Supplement_1/11543.full.pdf

Bergseth BJ, Gurney GG, Barnes ML et al. (2018) Addressing poaching in marine protected areas through voluntary surveillance and enforcement. Nature Sustainability 1: 421–426. https://doi.org/10.1038/s41893-018-0117-x

Bicknell AWJ, Godley BJ, Sheehan EV (2016) Camera technology for monitoring marine biodiversity and human impact. Frontiers in Ecology and the Environment 14(8): 424–432. https://doi.org/10.1002/fee.1322

Bizsel C, Yokes B, Pollard D et al. (2011) *Sciaena umbra. The IUCN Red List of Threatened Species* 2011: e.T198707A9080239. Accessed 08 June 2021

Boell SK, Cecez-Kecmanovic D (2015) On being 'systematic' in literature reviews in IS. Journal of Information Technology 30: 161–173. https://doi.org/10.1057/jit.2014.26

Borrini-Feyerabend G, Dudley N, Jaeger T et al. (2013) Governance of Protected Areas: From understanding to action. Best Practice Protected Area Guidelines. IUCN, Series No. 20, Gland, 124 pp. https://portals.iucn.org/library/sites/library/files/documents/PAG-020.pdf. Accessed 10 August 2021

Bowler DE, Bjorkman AD, Dornelas M et al. (2020) Mapping human pressures on biodiversity across the planet uncovers anthropogenic threat complexes. People and Nature 2 (2): 380–394. https://doi.org/10.1002/pan3.10071

Bruner AG, Gullison RE, Balmford A (2004) Financial Costs and Shortfalls of Managing and Expanding Protected-Area Systems in Developing Countries. BioScience 54 (12): 1119–1126. https://doi.org/10.1641/0006-3568(2004)054[1119:FCASOM]2.0.CO;2

Bruni CM, Fraser J, Schultz PW (2008) The Value of Zoo Experiences for Connecting People with Nature. Visitor Studies 11: 139–150. https://doi.org/10.1080/10645570802355489

Burns PB, Rohrich RD, Chung KC (2011) The levels of evidence and their role in evidence-based medicine. Plastic and Reconstructive Surgery 128(1): 305–310. https://doi.org/10.1097/PRS.0b013e318219c171

Butchart SHM, Walpole M, Collen B et al. (2010) Global biodiversity: indicators of recent declines. Science 328: 164–1168

Carabias J, Arriaga V, Cervantes-Gutiérrez V (2007) Public policies of environmental restoration in Mexico: limitations, progress, lags, and challenges. Botanical Sciences 80S: 85–100. https://doi.org/10.17129/botsci.1759

Carranza T, Manica A, Kapos V et al. (2014a) Mismatches between conservation outcomes and management evaluation in protected areas: A case study in the Brazilian Cerrado. Biological Conservation 173:10–16. https://doi.org/10.1016/j.biocon.2014.03.004

Carranza T, Balmford A, Kapos K et al. (2014b) Protected Area Effectiveness in Reducing Conversion in a Rapidly Vanishing Ecosystem: The Brazilian Cerrado. Conservation Letters: 7(3): 216–223. https://conbio.onlinelibrary.wiley.com/doi/full/10.1111/conl.12049

Cazalis V, Princé K, Mihoub JB et al. (2020) Effectiveness of protected areas in conserving tropical forest birds. Nature Communications 11: 4461. https://doi.org/10.1038/s41467-020-18230-0

CBD (1992) Text of the Convention. https://www.cbd.int/convention/text/. Accessed 11 March 2021

CBD (Convention on Biological Diversity) (1992) Text of the Convention. Available from: https://www.cbd.int/convention/text/

CBD (2010) Decision X/2: The Strategic Plan for Biodiversity 2011-2020 and the Aichi Biodiversity Targets. Adopted by the Conference of the Parties to the Convention on Biological Diversity at its Tenth Meeting. CBD, Nagoya. https://www.cbd.int/doc/decisions/cop-10/cop-10-dec-02-en.pdf. Accessed 11 March 2021

CBD (2015) Biodiversity and the 2030 Agenda for Sustainable Development. CBD, Montreal. https://www.cbd.int/development/doc/sdg-sep2015-flyer.pdf. Accessed 11 March 2021

CBD (2016) Biodiversity and Sustainable Development: Technical Note. UNEP/CBD/COP/13/10/Add.1. CBD, Montreal. https://www.cbd.int/doc/meetings/cop/cop-13/official/cop-13-10-add1-en.pdf. Accessed 11 March 2021

CBD (2018) Decision adopted by the Conference of the Parties to the Convention on Biological Diversity 14/8. Protected areas and other effective area-based conservation measures. CBD, Sharm El-Sheikh. https://www.cbd.int/doc/decisions/cop-14/cop-14-dec-08-en.pdf. Accessed 11 March 2021

CBD (2020) Global Biodiversity Outlook 5. CBD, Montreal. https://www.cbd.int/gbo/gbo5/publication/gbo-5-en.pdf. Accessed 15 July 2021

CBD (2021) First Detailed Draft of the new Post-2020 Global Biodiversity Framework. https://www.cbd.int/article/draft-1-global-biodiversity-framework. Accessed 12 August 2021

Ceballos G, García A, Ehrlich PR (2010) The sixth extinction crisis. Loss of animal populations and species. Journal of Cosmology 8: 1821–1831

Ceballos G, Ehrlich PR, Barnosky AD et al. (2015) Accelerated modern human–induced species losses: Entering the sixth mass extinction. Science Advances 1(5): e1400253. https://advances.sciencemag.org/content/1/5/e1400253

Ceccon E, Barrera-Cataño JI, Aronson J et al. (2015) The socioecological complexity of ecological restoration in Mexico. Restoration Ecology 23: 331–336. https://doi.org/10.1111/rec.12228

Chagos Conservation Trust (2021) *Enforcing the world's largest marine reserve*. https://chagos-trust.org/news/enforcing-the-worlds-largest-marine-reserve. Accessed 9 August 2021

Chandler M, See L, Copas K et al. (2017) Contribution of citizen science towards international biodiversity monitoring. Biological Conservation 213(B): 280–294. https://doi.org/10.1016/j.biocon.2016.09.004

Chuvieco E, Aguado I, Yebra M et al. (2010) Development of a framework for fire risk assessment using remote sensing and geographic information system technologies. Ecological Modelling 221: 46–58. https://doi.org/10.1016/j.ecolmodel.2008.11.017

Clayton S, Myers G (2015) Conservation Psychology: Understanding and Promoting Human Care for Nature. John Wiley & Sons, 2nd ed.

Coad L, Leverington F, Knights K et al (2015) Measuring impact of protected area management interventions: current and future use of the Global Database of Protected Area Management Effectiveness. Philosophical Transactions of The Royal Society B 370(1681). https://royalsocietypublishing.org/doi/10.1098/rstb.2014.0281

Coates DJ, Byrne M, Moritz C (2018) Genetic Diversity and Conservation Units: Dealing With the Species-Population Continuum in the Age of Genomics. Frontiers in Ecology and Evolution 6:165. https://doi.org/10.3389/fevo.2018.00165

Coetzee BWT, Gaston KJ, Chown SL (2014) Local Scale Comparisons of Biodiversity as a Test for Global Protected Area Ecological Performance: A Meta-Analysis. PLoS ONE 9(8): e105824. https://doi.org/10.1371/journal.pone.0105824

Coldrey KM, Turpei JK (2021) The future representativeness of Madagascar's protected area network in the face of climate change. African Journal of Ecology 59(1): 253–263. https://doi.org/10.1111/aje.12819

Collen B, McRae L, Nicholson E (2016) Species propulation trends in protected areas. In: Joppa LN, Baillie JEM, Robinson JG (eds) Protected areas. Are they safeguarding biodiversity? Wiley, Chichester, pp. 105–212.

Collins C, Nuno A, Broderick A et al. (2021) Understanding persistent non-compliance in a remote, large-scale marine protected area. Frontiers in Marine Science 8: 8:650276. https://doi.org/10.3389/fmars.2021.650276

Comunidad de Madrid (1997) Plan Rector de Uso y Gestión del Parque Regional de la Cuenca Alta del Manzanares. Madrid. https://www.comunidad.madrid/transparencia/sites/default/files/plan/document/601_821_mgr_cit_13710_prcam_prug_0.pdf. Accessed 15 July 2021

CONANP (2018) Programa de Conservación de Especies en Riesgo (PROCER). Comisión Nacional de Áreas Naturales Protegidas. https://www.gob.mx/conanp/acciones-y-programas/programa-de-conservacion-de-especies-en-riesgo

Conde DA, Flesness N, Colchero F et al.(2011) An Emerging Role of Zoos to Conserve Biodiversity. Science 331 (6023): 1390–1391. https://doi.org/10.1126/science.1200674

Cook CN, Carter RW (Bill), Hockings M (2014) Measuring the accuracy of management effectiveness evaluations of protected areas. Journal of Environmental Management 139: 164–171. https://doi.org/10.1016/j.jenvman.2014.02.023

Cook CN, Valkan RS, McGeoch MA (2019) Beyond total area protected: A new set of metrics to measure progress in building a robust protected area estate. Global Environmental Change 58: 101963. https://doi.org/10.1016/j.gloenvcha.2019.101963

Costello MJ (2015) Biodiversity: the known, unknown and rates of extinction. Current Biology 25(9): R368–R371. https://doi.org/10.1016/j.cub.2015.03.051

Costanza R, de Groot R, Sutton P, van der Ploeg S, Anderson SJ, Kubiszewski I, Farber S, Turner K (2014) Changes in the global value of ecosystem services. Global Environmental Change 26: 152–158

Council of Europe (2021) *Emerald Network of Areas of Special Conservation Interest*. https://www.coe.int/en/web/bern-convention/emerald-network. Accessed 11 March 2021

Craigie ID, Baillie JEM, Balmford A et al (2010) Large mammal population declines in Africa's protected areas. Biological Conservation 143(9): 2221–2228 https://doi.org/10.1016/j.biocon.2010.06.007

Crook M, Short D, South N (2018) Ecocide, genocide, capitalism and colonialism: Consequences for indigenous peoples and glocal ecosystems environments. Theoretical Criminology 22(3): 298–317. https://doi.org/10.1177/1362480618787176

Crutzen P (2002) Geology of mankind. Nature 415: 23. https://doi.org/10.1038/415023a

Daily G (1997) Nature's services: societal dependence on natural ecosystems. Island Press, Washington, DC

Dalton T, Jin D (2010) Extent and frequency of vessel oil spills in US marine protected areas. Marine Pollution Bulletin 60: 1939–1945. https://doi.org/10.1016/j.marpolbul.2010.07.036

De la Varga Pastor A, Pons Solé J (2018) Innovative legal tools applied in land stewardship for the conservation of ecosystem services in Catalonia. Ecosystem Services 29 (B): 395–403. https://doi.org/10.1016/j.ecoser.2017.08.010

de Oliveira Soares M, Costa Lucas C (2018) Towards large and remote protected areas in the South Atlantic Ocean: St. Peter and St. Paul's Archipelago and the Vitória-Trindade Seamount Chain. Marine Policy 93: 101–103. https://doi.org/10.1016/j.marpol.2018.04.004

Dharmaratne GS, Yee Sang F, Walling LJ (2000) Tourism potentials for financing protected areas. Annals of Tourism Research 27: 590–610. https://doi.org/10.1016/S0160-7383(99)00109-7

Dias MS, Tedesco PA, Hugueny B et al. (2017) Anthropogenic stressors and riverine fish extinctions. Ecological Indicators 79: 37–46. https://doi.org/10.1016/j.ecolind.2017.03.053

Dirzo R, Young HS, Galetti M et al. (2014) Defaunation in the Anthropocene. Science 345(6195): 401–406. https://doi.org/10.1126/science.1251817

Dudley, N. (Editor) (2008). Guidelines for Applying Protected Area Management Categories. IUCN, Gland, 86 pp. https://www.iucn.org/sites/dev/files/import/downloads/iucn_assignment_1.pdf. Accessed 15 July 2021

Dudley N, Jonas H, Nelson F et al (2018) The essential role of other effective area-based conservation measures in achieving big bold conservation targets. Global Ecology and Conservation 15: e00424. https://doi.org/10.1016/j.gecco.2018.e00424

Duro DC, Coops NC, Wulder MA (2007) Development of a large area biodiversity monitoring system driven by remote sensing. Progress in Physical Geography: Earth and Environment 31(3): 235–260. https://doi.org/10.1177/0309133307079054

EAZA (2021) EAZA Ex situ Programmes. https://www.eaza.net/conservation/programmes/. Accessed 11 Mars 2021

Ebrahimi E, Sayahnia R, Ranjbaran Y et al. (2021) Dynamics of threatened mammalian distribution in Iran's protected areas under climate change. Mamm Biol. https://doi.org/10.1007/s42991-021-00136-z

EEC (1992) Council Directive 92/43/EEC of 21 May 1992 on the conservation of natural habitats and of wild fauna and flora. https://eur-lex.europa.eu/legal-content/EN/TXT/?uri=celex%3A31992L0043. Accessed 27 September 2021

Edgar G, Stuart-Smith R, Willis T et al. (2014) Global conservation outcomes depend on marine protected areas with five key features. Nature 506: 216–220. https://doi.org/10.1038/nature13022

Egerton JP, Johnson AF, Turner J et al. (2018) Hydroacoustics as a tool to examine the effects of Marine Protected Areas and habitat type on marine fish communities. Scientific Reports 8: 47. https://doi.org/10.1038/s41598-017-18353-3

Eklund J, Coad L, Geldmann J et al. (2019) What constitutes a useful measure of protected area effectiveness? A case study of management inputs and protected area impacts in Madagascar. Conservation Science and Practice 1: e107. https://doi.org/10.1111/csp2.107

Ervin J (2003) Rapid Assessment and Prioritization of Protected Area Management (RAPPAM) Methodology. WWF, Gland

European Commission (2020a) Natura 2000. State of Nature in the EU. n° 49. Directorate-General for Environment (European Commission). https://op.europa.eu/en/publication-detail/-/publication/c4622350-56e4-11eb-b59f-01aa75ed71a1/language-en/format-PDF/source-187697953. Accessed 11 March 2021

European Commission (2020b) Biodiversity strategy for 2030. https://ec.europa.eu/environment/strategy/biodiversity-strategy-2030_en. Accessed 29 June 2021

European Commission (2021a) The Habitats Directive. https://ec.europa.eu/environment/nature/legislation/habitatsdirective/. Accessed 11 March 2021

European Commission (2021b) Data quality and completeness scoring in Article 17. https://tableau.discomap.eea.europa.eu/t/Natureonline/views/DataqualitycompletenessforArticle17/Article17Dataqualitycompleteness?:showAppBanner=false&:display_count=n&:showVizHome=n&:origin=viz_share_link&:isGuestRedirectFromVizportal=y&:embed=y. Accessed 10 August 2021

European Economic Community (1992) Council Directive 92/43/EEC of 21 May 1992 on the conservation of natural habitats and of wild fauna and flora. https://eur-lex.europa.eu/legal-content/EN/TXT/?uri=celex%3A31992L0043. Accessed 27 September 2021

European Environment Agency (2021) Copernicus Land Monitoring Service. Land use cases. https://land.copernicus.eu/user-corner/land-use-cases. Accessed 28 June 2021

FAO (2021) REDD+ Reducing Emissions from Deforestation and Forest Degradation. Accessed 28 June 2021

Ferrante L, Fearnside P (2019) Brazil's new president and 'ruralists' threaten Amazonia's environment, traditional peoples and the global climate. Environmental Conservation 46(4): 261–263. https://doi.org/10.1017/S0376892919000213

Fraser RH, Olthof I, Pouliot D (2009) Monitoring land cover change and ecological integrity in Canada's national parks. Remote Sensing of Environment 113(7): 1397–1409. https://doi.org/10.1016/j.rse.2008.06.019

Fuller R, McDonald-Madden E, Wilson K et al. (2010) Replacing underperforming protected areas achieves better conservation outcomes. Nature 466: 365–367. https://doi.org/10.1038/nature09180

Gaffney O, Steffen W (2017) The Anthropocene equation. The Anthropocene Review 4 (1): 53–61. https://doi.org/10.1177/2053019616688022

GANP, Global Alliance of National Parks (2021) *Korup National Park*.https://national-parks.org/cameroon/korup#&gid=null&pid=3. Accessed 18 June 2021

Gardner CJ, Nicoll ME, Birkinshaw C et al. (2018) The rapid expansion of Madagascar's protected area system. Biological Conservation 220: 29–36. https://doi.org/10.1016/j.biocon.2018.02.011

Gaston KJ, Jackson SF, Cantú-Salazar L et al. (2008) The Ecological Performance of Protected Areas. Annual Review of Ecology, Evolution, and Systematics 39: 93–113. https://doi.org/10.1146/annurev.ecolsys.39.110707.173529

Geldmann G, Barnes M, Coad L et al. (2013) Effectiveness of terrestrial protected areas in reducing habitat loss and population declines. Biological Conservation 161: 230–238. https://doi.org/10.1016/j.biocon.2013.02.018

Geldmann J, Coad L, Barnes M et al. (2015) Changes in protected area management effectiveness over time: A global analysis. Biological Conservation 191: 692–699. https://doi.org/10.1016/j.biocon.2015.08.029

Geldmann J, Manica A, Burgess ND et al (2019) A global-level assessment of the effectiveness of protected areas at resisting anthropogenic pressures. PNAS 116 (46): 23209–23215. https://doi.org/10.1073/pnas.1908221116

Geldmann J, Manica A, Burgess, ND et al., (2020) A global-level assessment of the effectiveness of protected areas at resisting anthropogenic pressures. PNAS 116(46): 23209–23215.

Giakoumi S, McGowan J, Mills M et al. (2018) Revisiting "Success" and "Failure" of Marine Protected Areas: A Conservation Scientist Perspective. Frontiers in Marine Science 5: 223. https://www.frontiersin.org/article/10.3389/fmars.2018.00223

Gillingham PK, Bradbury RB, Roy DB et al. (2015) The effectiveness of protected areas in the conservation of species with changing geographical ranges. Biological Journal of the Linnean Society 115(3): 707–717. https://doi.org/10.1111/bij.12506

Global Conservation (2018) Over One Thousand Park Rangers Die in 10 Years Protecting Our Parks and Wildlife. https://globalconservation.org/news/over-one-thousand-park-rangers-die-10-years-protecting-our-parks/. Accessed 10 August 2021

Grosberg RK, Vermeij GJ, Wainwright PC (2012) Biodiversity in water and on land. Current Biology 22(21): R900–R903. https://doi.org/10.1016/j.cub.2012.09.050

Grubb P, Butynski TM, Oates JF et al. (2003) Assessment of the diversity of African primates. International Journal of Primatology 24: 1301–1357. https://doi.org/10.1023/B:IJOP.0000005994.86792.b9

Gusset M, Dick G (2011) The global reach of zoos and aquariums in visitor numbers and conservation expenditures. Zoobiology 30(5): 566–569. https://doi.org/10.1002/zoo.20369

Halpern BS, Walbridge S, Selkoe KA et al. (2008) A Global Map of Human Impact on Marine Ecosystems. Science 319(5865): 948–952. https://doi.org/10.1126/science.1149345

Harmelin-Vivien M, Cottalorda JM, Dominici JM et al. (2015) Effects of reserve protection level on the vulnerable fish species Sciaena umbra and implications for fishing management and policy. Global Ecology and Conservation 3: 279–287. https://doi.org/10.1016/j.gecco.2014.12.005

Hellwig N, Walz A, Markovic D (2019) Climatic and socioeconomic effects on land cover changes across Europe: Does protected area designation matter? PLoS ONE 14(7): e0219374. https://doi.org/10.1371/journal.pone.0219374

Hendry AP, Lohmann LG, Conti E et al. (2010) Evolutionary Biology in Biodiversity Science, Conservation, and Policy: A Call to Action. Evolution 64: 1517–1528. https://doi.org/10.1111/j.1558-5646.2010.00947.x

Herrick JE, Schuman GE, Rango A (2006) Monitoring ecological processes for restoration projects. Journal for Nature Conservation 14: 161–171. https://doi.org/10.1016/j.jnc.2006.05.001

Hill NA, Barrett N, Lawrence E et al. (2014) Quantifying Fish Assemblages in Large, Offshore Marine Protected Areas: An Australian Case Study. PLoS ONE 9(10): e110831. https://doi.org/10.1371/journal.pone.0110831

Hockings M, Stolton S, Dudley N (2000) Evaluating effectiveness: a framework for assessing the management of protected areas. IUCN, Gland

Hockings M, Stolton S, Leverington F et al (2006) Evaluating Effectiveness: a framework for assessing management effectiveness of protected areas. IUCN, Gland-Cambridge

Hoffmann AA, Sgrò CM, Kristensen TN (2017) Revisiting Adaptive Potential, Population Size, and Conservation. Trends in Ecology & Evolution 32(7): 506–517. https://doi.org/10.1016/j.tree.2017.03.012

Hoffman M, Koenig K, Bunting G et al. (2016) Biodiversity Hotspots (version 2016.1) (2016.1) [Data set]. Zenodo. https://doi.org/10.5281/zenodo.3261807

Infield M (1988) Hunting, Trapping and Fishing in Villages within and on the Periphery of the Korup National Park. Report to World Wildlife Fund, Gland.

IPBES (2019) Global assessment report on biodiversity and ecosystem services of the Intergovernmental Science-Policy Platform on Biodiversity and Ecosystem Services. Brondizio

ES, Settele J, Díaz S et al. (eds). IPBES secretariat, Bonn, 1148 pages. https://doi.org/10.5281/zenodo.3831673. Accessed 16 May 2021

IUCN (2012) IUCN Red List categories and criteria, version 3.1, second edition. IUCN, Gland and Cambridge. https://portals.iucn.org/library/node/10315. Accessed 11 March 2021

IUCN (2016) A Global Standard for the Identification of Key Biodiversity Areas. IUCN. Gland and Cambridge. https://portals.iucn.org/library/sites/library/files/documents/2016-048.pdf. Accessed 18 March 2021

IUCN (2017) IUCN Green List of Protected and Conserved Areas: Standard, Version 1.1. IUCN, Gland

IUCN-WCPA (2019) Recognising and reporting other effective area-based conservation measures. ICN, Gland. https://portals.iucn.org/library/sites/library/files/documents/PATRS-003-En.pdf. Accessed 11 March 2021

IUCN (2020) Key results of the IUCN World Heritage Outlook. IUCN. Gland and Cambridge. https://worldheritageoutlook.iucn.org/results. Accessed 23 June 2021

IUCN (2021a) IUCN Green List of Protected and Conserved Areas. IUCN. Gland and Cambridge. https://www.iucn.org/theme/protected-areas/our-work/iucn-green-list-protected-and-conserved-areas. Accessed 12 August 2021

IUCN (2021b) The IUCN Red List of Threatened Species. IUCN. Gland and Cambridge. http://www.iucnredlist.org. Accessed 16 July 2021

Jacobson AP, Riggio J, Tait AM. et al. (2019) Global areas of low human impact ('Low Impact Areas') and fragmentation of the natural world. Scientific Reports 9, 14179. https://doi.org/10.1038/s41598-019-50558-6

Johnson CN, Balmford A, Brook BW et al. (2017) Biodiversity Losses and Conservation Responses in the Anthropocene. Science 356 (6335): 270-275. https://doi.org/10.1126/science.aam9317

Jones KR, Klein CJ, Halpern BS et al. (2018) The Location and Protection Status of Earth's Diminishing Marine Wilderness. Current Biology 28(15): P2506–2512.e3. https://doi.org/10.1016/j.cub.2018.06.010

Joppa L, Pfaff A (2010) Reassessing the forest impacts of protection The challenge of nonrandom location and a corrective method. Annals of the New York Academy Sciences 1185: 135–149. https://doi.org/10.1111/j.1749-6632.2009.05162.x

Just J, Kristensen RM, Olesen J (2014) *Dendrogramma*, New Genus, with Two New Non-Bilaterian Species from the Marine Bathyal of Southeastern Australia (Animalia, Metazoa *incertae sedis*) – with Similarities to Some Medusoids from the Precambrian Ediacara. PLoS ONE, 9(9): e102976. https://doi.org/10.1371/journal.pone.0102976

Kalkandelen K, O'Byrne D (2017) On ecocide: toward a conceptual framework. Journal of Social Theory 18(3): 333–349. https://doi.org/10.1080/1600910X.2017.1331857

Karanth KK, Nepal SK (2012) Local Residents Perception of Benefits and Losses From Protected Areas in India and Nepal. Environmental Management 49: 372–386. https://doi.org/10.1007/s00267-011-9778-1

Kearney S, Adams V, Fuller R et al. (2020) Estimating the benefit of well-managed protected areas for threatened species conservation. Oryx 54(2): 276–284. https://doi.org/10.1017/S0030605317001739

Keulartz J (2012) The Emergence of Enlightened Anthropocentrism in Ecological Restoration. Nature and Culture 7(1). https://doi.org/10.3167/nc.2012.070104

Klink CA, Moreira AG (2002) Past and current human occupation, and land use. In: Oliveira PS, Marquis RJ (eds). The cerrados of Brazil: ecology and natural history of a neotropical savanna. Columbia University Press, New York, pp. 69–88

Klink CA, Machado RB (2005) Conservation of the Brazilian cerrado. Conservation Biology 19: 707–713. https://www.jstor.org/stable/3591058

Krausmann F, Erb KH, Gingrich S et al. (2013) Global human appropriation of net primary production doubled in the 20th century. PNAS 110(25): 10324–10329. https://doi.org/10.1073/pnas.1211349110

Laikre L (2010) Genetic diversity is overlooked in international conservation policy implementation. Conservation Genetics 11: 349–354. https://doi.org/10.1007/s10592-009-0037-4

Lajeunesse A (2021) Canada's Arctic Offshore and Patrol Ships (AOPS): Their history and purpose. Marine Policy 124: 104323. https://doi.org/10.1016/j.marpol.2020.104323

Lehikoinen P, Tiusanen M, Santangeli A. et al. (2021) Increasing protected area coverage mitigates climate-driven community changes. Biological Conservation 253: 108892. https://doi.org/10.1016/j.biocon.2020.108892

Leigh DM, Hendry A, Vazquez-Dominguez E et al. (2018) Six Percent Loss of Genetic Variation in Wild Populations since the Industrial Revolution. BioRxiv 488650. https://doi.org/10.1101/488650

Leverington F, Costa KL, Pavese H et al (2010) A Global Analysis of Protected Area Management Effectiveness. Environmental Management 46: 685–698. https://doi.org/10.1007/s00267-010-9564-5

Leverington F, Costa KL, Pavese H et al. (2010a) A Global Analysis of Protected Area Management Effectiveness. Environmental Management 46: 685–698. https://doi.org/10.1007/s00267-010-9564-5

Leverington F, Lemos K, Courrau J et al. (2010b) Management effectiveness evaluation in protected areas – a global study. Second Edition. Brisbane, University of Queensland.

Linder JM, Oates JF (2011) Differential impact of bushmeat hunting on monkey species and implications for primate conservation in Korup National Park, Cameroon. Biological Conservation 144: 738–745. https://doi.org/10.1016/j.biocon.2010.10.023

Lindsey P, Baghai M, Bigurube G et al. (2021) Attracting investment for Africa's protected areas by creating enabling environments for collaborative management partnerships. Biological Conservation 255, 108979. https://doi.org/10.1016/j.biocon.2021.108979

Liu J, Linderman M, Ouyang Z et al. (2001). Ecological Degradation in Protected Areas: The Case of Wolong Nature Reserve for Giant Pandas. Science 292 (5514): 98–101. https://doi.org/10.1126/science.1058104

Lockwood M (2010) Good governance for terrestrial protected areas: A framework, principles and performance outcomes. Journal of Environmental Management 91 (3): 754–766. https://doi.org/10.1016/j.jenvman.2009.10.005

Loiseau C, Thiault L, Devillers R et al. (2021) Cumulative impact assessments highlight the benefits of integrating land-based management with marine spatial planning. Science of The Total Environment 787: 147339. https://doi.org/10.1016/j.scitotenv.2021.147339

Lyons KS, Amatangelo, K, Behrensmeyer, A et al. (2016) Holocene shifts in the assembly of plant and animal communities implicate human impacts. Nature 529: 80–83.

Mammides C (2020) A global assessment of the human pressure on the world's lakes. Global Environmental Change 63: 102084. https://doi.org/10.1016/j.gloenvcha.2020.102084

Margulis L, Chapman MJ (2009) KINGDOM PROKARYOTAE (Bacteria, Monera, Prokarya). In: L Margulis and MJ Chapman (Eds.) Kingdoms and Domains. An Illustrated Guide to the Phyla of Life on Earth. Academic Press, p 35–107. https://doi.org/10.1016/B978-0-12-373621-5.00001-5

Marten G (2005) Marine Sanctuary: Restoring a Coral-Reef Fishery (Apo Island, Philippines). http://www.ecotippingpoints.org/our-stories/indepth/philippines-apo-marine-sanctuary-coral-reef-fishery.html. Accessed 11 June 2021

Mascia MB, Pailler S (2011) Protected area downgrading, downsizing, and degazettement (PADDD) and its conservation implications. Conservation Letters 4: 9–20. https://doi.org/10.1111/j.1755-263X.2010.00147.x

Massaro M, Dumay J, Guthrie J (2016) On the shoulders of giants: undertaking a structured literature review in accounting. Accounting, Auditing & Accountability Journal 29(5): 767–801. https://doi.org/10.1108/AAAJ-01-2015-1939

Mayr E (1942) Systematics and the Origin of Species. Columbia University Press, New York

MEA (Millenium Ecosystem Assessment) (2005) Ecosystems and well-being. Synthesis. Island Press. Washington DC. Available from: https://www.millenniumassessment.org/en/index.html

Mendoza JC, Clemente S, Hernández JC (2020) Modeling the role of marine protected areas on the recovery of shallow rocky reef ecosystem after a catastrophic submarine volcanic eruption. Marine Environmental Research 155: 104877. https://doi.org/10.1016/j.marenvres.2020.104877

Méral P, Froger G, Andriamahefazafy F (2011). Financing Protected Areas in Madagascar: New Methods. In: Rodary E, Aubertin C (eds). Protected Areas, Sustainable Land? 1st edn. Routledge, London, p 85–101. https://doi.org/10.4324/9781315602646

Meyer C, Kreft H, Guralnick R et al. (2015) Global priorities for an effective information basis of biodiversity distributions. Nature Communications 6: 8221. https://doi.org/10.1038/ncomms9221

Miller DGM, Slicer NM, Hanich Q (2013) Monitoring, control and surveillance of protected areas and specially managed areas in the marine domain. Marine Policy 39: 64–71. https://doi.org/10.1016/j.marpol.2012.10.004

Mingarro M, Lobo JM (2021) Connecting protected areas in the Iberian peninsula to facilitate climate change tracking. Environmental Conservation: 1–10. https://doi.org/10.1017/S037689292100014X

Minin ED, Toivonen T (2015) Global Protected Area Expansion: Creating More than Paper Parks. BioScience 65(7): 637–638. https://doi.org/10.1093/biosci/biv064

Mora C, Sale PF (2011) Ongoing global biodiversity loss and the need to move beyond protected areas: a review of the technical and practical shortcomings of protected areas on land and sea. Marine Ecology Progress Series 434: 251–266. https://doi.org/10.3354/meps09214

Mounce R, Smith P, Brockington S (2017) Ex situ conservation of plant diversity in the world's botanic gardens. Nature Plants 3: 795–802. https://doi.org/10.1038/s41477-017-0019-3

Myers N, Mittermeier RA, Mittermeier CG et al. (2000) Biodiversity hotspots for conservation priorities. Nature 403: 853–858. https://www.nature.com/articles/35002501

Nagendra H (2008) Do Parks Work? Impact of Protected Areas on Land Cover Clearing. Ambio 37(5): 330–337. https://doi.org/10.1579/06-R-184.1

Naidoo R, Adamowicz WL (2005) Economic benefits of biodiversity exceed costs of conservation at an African rainforest reserve. PNAS 102(46): 16712–16716. https://doi.org/10.1073/pnas.0508036102

Nchanji YK, Ramcilovic-Suominen S, Kotilainen J (2021) Power imbalances, social inequalities and gender roles as barriers to true participation in national park management: The case of Korup National Park, Cameroon. Forest Policy and Economics 130: 102527. https://doi.org/10.1016/j.forpol.2021.102527

Niner H, Ardron JA, Escobar EG et al. (2018) Deep-Sea Mining With No Net Loss of Biodiversity—An Impossible Aim. Frontiers in Marine Science 5. https://doi.org/10.3389/fmars.2018.00053

Nolte C, Agrawal A (2013) Linking Management Effectiveness Indicators to Observed Effects of Protected Areas on Fire Occurrence in the Amazon Rainforest. Conservation Biology 27(1): 155–165. https://doi.org/10.1111/j.1523-1739.2012.01930.x

Noss RF, Dobson AP, Baldwin R et al. (2012) Bolder thinking for conservation. Conservation Biology 26(1): 1–4. https://doi.org/10.1111/j.1523-1739.2011.01738.x

O'Regan SM, Archer SK, Friesen SK et al. (2021) A Global Assessment of Climate Change Adaptation in Marine Protected Area Management Plans. Frontiers in Marine Science 8. https://www.frontiersin.org/article/10.3389/fmars.2021.711085

Oates JF, Bergl RA, Linder JM (2004) Africa's Gulf of Guinea Forests: Biodiversity Patterns and Conservation Priorities. Conservation International Center for Applied Biodiversity Science. https://bioone.org/ebooks/. Accessed 19 July 2021

OECD (2021) Monitoring land cover change. https://www.oecd.org/environment/indicators-modelling-outlooks/monitoring-land-cover-change.htm. Accessed 28 June 2021

Okoli C, Schabram K (2010) A Guide to conducting a Systematic Literature Review of Information Systems Research. Working Papers on Information Systems 10(26). https://ssrn.com/abstract=1954824 or https://doi.org/10.2139/ssrn.1954824. Accessed 15 April 2021

Oliveira U, Soares-Filho BS, Paglia AP et al. (2017) Biodiversity conservation gaps in the Brazilian protected areas. Scientific Reports 7: 9141. https://doi.org/10.1038/s41598-017-08707-2

Palmer MA, Menninger HL, Bernhardt E (2010) River restoration, habitat heterogeneity and biodiversity: a failure of theory or practice? Freshwater Biology 55(1): 205–222. https://doi.org/10.1111/j.1365-2427.2009.02372.x

Parcs Victoria (2021) Research Partners Programme. https://www.parks.vic.gov.au/get-into-nature/conservation-and-science/science-and-research/research-partners-program. Accessed 27 August 2021

Pereira HM, Navarro LM, Martins IS (2012) Global biodiversity change: the bad, the good, and the unknown. Annual Review of Environment and Resources 37: 25–50. https://doi.org/10.1146/annurev-environ-042911-093511

Perry AL, Blanco J, Fournier N et al. (2020) Unmanaged = Unprotected: Europe's marine paper parks. Oceana, Brussels. 52 pp. https://doi.org/10.5281/zenodo.4312546

Pfeifer M, Burgess ND, Swetnam RD et al (2012) Protected areas: mixed success in conserving east Africa's evergreen forests. PLoS One 7 (6), e39337. https://doi.org/10.1371/journal.pone.0039337

Phillipe H, Brinkmann H, Copley RR et al. (2011) Acoelomorph flatworms are deuterostomes related to *Xenoturbella*. Nature 470: 255–258. https://doi.org/10.1038/nature09676

Potapov P, Hansen MC, Laestadius L et al. (2017) The last frontiers of wilderness: Tracking loss of intact forest landscapes from 2000 to 2013. Science Advances 3(1): e1600821. https://doi.org/10.1126/sciadv.1600821

Protected Planet (2020) Protected Planet Report 2020. Coverage. Chapter 3 https://livereport.protectedplanet.net/chapter-3 Accessed 18 March 2021. Accessed 27 August 2021

Protected Planet (2021) Management effectiveness (PAME) https://www.protectedplanet.net/en/thematic-areas/protected-areas-management-effectiveness-pame?tab=Results. Accessed 27 August 2021

Pullin AS (2002) Conservation Biology. Cambridge University Press. Cambridge, UK

Purvis A, Hector A (2000) Getting the measure of biodiversity. Nature 405: 212–219

Pyhälä A, Eklund J, McBride MF et al. (2019) Managers' perceptions of protected area outcomes in Madagascar highlight the need for species monitoring and knowledge transfer. Conservation, Science and Practice 1(2): e6. https://doi.org/10.1111/csp2.6

Qin S, Golden Kroner R E, Cook C et al. (2019) Protected area downgrading, downsizing, and degazettement as a threat to iconic protected areas. Conservation Biology: The Journal of the Society for Conservation Biology 33(6): 1275–1285. https://doi.org/10.1111/cobi.13365

Ratter JA, Ribeiros JF, Bridgewater S (1997) The Brazilian Cerrado Vegetation and Threats to its Biodiversity. Annals of Botany 80: 223–230. https://doi.org/10.1006/anbo.1997.0469

Rija AA (2017) Spatial Pattern of Illegal Activities and the Impact on Wildlife Populations in Protected Areas in the Serengeti Ecosystem. PhD thesis, University of York.

Rodríguez JP and Rodríguez-Clark KM (2001) Even 'paper parks' are important. Trends in Ecology and Evolution 16(1): 17. https://doi.org/10.1016/S0169-5347(00)02050-4

Rodríguez-Rodríguez D (2012) New issues on protected area management. In: Sladonja B (ed) Protected Area Management. IntechOpen, Rijeka, p. 19–42

Rodríguez-Rodríguez D, Martínez-Vega J (2012) Proposal of a system for the integrated and comparative assessment of protected areas. Ecological Indicators 23: 566–572. https://doi.org/10.1016/j.ecolind.2012.05.009

Rodríguez-Rodríguez D, Sciberras M, Foster NL et al. (2015a) Status of management effort in 153 marine protected areas across the English Channel. Marine Pollution Bulletin 94(1–2): 168–175. https://doi.org/10.1016/j.marpolbul.2015.02.036

Rodríguez-Rodríguez D, Sánchez-Espinosa A, Schröder C et al. (2015b) Cumulative pressures and low protection: a concerning blend for Mediterranean MPAs. Marine Pollution Bulletin 101(1): 288–295. https://doi.org/10.1016/j.marpolbul.2015.09.039

Rodríguez-Rodríguez D, Rodríguez J, Abdul Malak D (2016) Development and testing of a new framework for rapidly assessing legal and managerial protection afforded by marine protected areas: Mediterranean Sea case study. Journal of Environmental Management 167: 29–37. https://doi.org/10.1016/j.jenvman.2015.11.016

Rodríguez-Rodríguez D, Martínez-Vega J (2018) Protected area effectiveness against land development in Spain. Journal of Environmental Management 215: 345–357. https://doi.org/10.1016/j.jenvman.2018.03.011

Rodríguez-Rodríguez D, Martínez-Vega J (2018a) Representation and protection of threatened biodiversity by the largest Spanish regional network of protected areas. Endangered Species Research 35:125–139. https://doi.org/10.3354/esr00878

Rodríguez-Rodríguez D, Martínez-Vega J (2018b) Protected area effectiveness against land development in Spain. Journal of Environmental Management 215: 345–357. https://doi.org/10.1016/j.jenvman.2018.03.011

Russ GR, Lowe JR, Rizzari JR et al. (2017) Partitioning no-take marine reserve (NTMR) and benthic habitat effects on density of small and large-bodied tropical wrasses. PLoS ONE 12(12): e0188515. https://doi.org/10.1371/journal.pone.0188515

Sanderson EW, Jaiteh M, Levy MA et al. (2002) The Human Footprint and the Last of the Wild. BioScience 52(10): 891–904. https://doi.org/10.1641/0006-3568(2002)052[0891:THFATL]2.0.CO;2

Sano EE, Rosa R, Brito JLS et al. (2010) Mapeamento do uso do solo e cobertura vegetal—bioma Cerrado: ano base 2002. MMA/SBF, Brasilia. https://www.terrabrasilis.org.br/ecotecadigital/pdf/serie-biodiversidade-36-mapeamento-do-uso-do-solo-e-cobertura-vegetal-bioma-cerrado-ano-base-2002-.pdf. Accessed 6 July 2021

Santini L, Boitani L, Maiorano L et al. (2016) Effectiveness of protected areas in conserving large carnivores in Europe. In: Joppa LN, Baillie JEM, Robinson JG (eds) Protected areas. Are they safeguarding biodiversity? Wiley, Chichester, pp. 122–133.

Schleicher J, Peres CA, Leader-Williams N (2019) Conservation performance of tropical protected areas: How important is management? Conservation Letters 12:e12650. https://doi.org/10.1111/conl.12650

Schmitz MF, Matos DGG, De Aranzabal I et al. (2012) Effects of a protected area on land-use dynamics and socioeconomic development of local populations. Biological Conservation 149 (1): 122–135. https://doi.org/10.1016/j.biocon.2012.01.043

Schoneveld-de Lange N, Meijaard E, Löhr A (2016) South to south learning in great ape conservation. American Journal of Primatology 78 (6): 669–678. https://doi.org/10.1002/ajp.22531

Sekhar N (1998) Crop and livestock depredation caused by wild animals in protected areas: The case of Sariska Tiger Reserve, Rajasthan, India. Environmental Conservation 25(2): 160–171. https://doi.org/10.1017/S0376892998000204

SER (2021) What is Ecological Restoration? https://www.ser-rrc.org/what-is-ecological-restoration/. Accessed 11 March 2021

Singh SK, Srivastava PK, Gupta M et al. (2014) Appraisal of land use/land cover of mangrove forest ecosystem using support vector machine. Environmental Earth Science 71: 2245–2255. https://doi.org/10.1007/s12665-013-2628-0

Soykan CU, Lewison RL (2015) Using community-level metrics to monitor the effects of marine protected areas on biodiversity. Conservation Biology 29(3): 775–83. https://doi.org/10.1111/cobi.12445

Spracklen BD, Kalamandeen M, Galbraith D et al (2015) A Global Analysis of Deforestation in Moist Tropical Forest Protected Areas. PLoS ONE 10(12): e0143886. https://doi.org/10.1371/journal.pone.0143886

Squires DF (1959) Corals and coral reefs in the Gulf of California. Bulletin of the American Museum of Natural History 118: 378–384. https://digitallibrary.amnh.org/handle/2246/1267

Staples M, Niazi M (2007) Experiences using systematic review guidelines. Journal of Systems and Software 80(9): 1425–1437. https://doi.org/10.1016/j.jss.2006.09.046

Steffen W, Richardson K, Rockström J et al. (2015) Planetary boundaries: Guiding human development on a changing planet. Science 347(6223): 1259855. https://doi.org/10.1126/science.1259855

Stolton S, Hockings M, Dudley N et al. (2007) Reporting progress in protected areas: a site-level Management Effectiveness Tracking Tool, 2nd edn. http://assets.panda.org/downloads/mett2_final_version_july_2007.pdf. Accessed 18 December 2009

Stop Ecocide Foundation (2021) Independent Expert Panel for the Legal Definition of Ecocide. Commentary and Core Text. https://static1.squarespace.com/static/5ca2608ab914493c64ef1f6d/t/60d1e6e604fae2201d03407f/1624368879048/SE+Foundation+Commentary+and+core+text+rev+6.pdf. Accessed 29 June 2021

Sun CZ, Wang S, Zou W (2016) Chinese marine ecosystem services value: regional and structural equilibrium analysis. Ocean Coastal & Management 125: 70–83. https://doi.org/10.1016/j.ocecoaman.2016.03.001

Svalbard Global Seed Vault (2021) Safeguarding Seeds for the Future. https://www.seedvault.no/. Accessed 11 March 2021

Sweke EA, Assam JM, Chande AI et al. (2016) Comparing the Performance of Protected and Unprotected Areas in Conserving Freshwater Fish Abundance and Biodiversity in Lake Tanganyika, Tanzania. International Journal of Ecology 2016: 7139689. https://doi.org/10.1155/2016/7139689

Tanzania National Parks (2021) Mahale Mountains National Park. Nature. https://web.archive.org/web/20090402124005/http://www.mahalepark.org/nature.html. Accessed 8 July 2021

TEEB (2010). The Economics of Ecosystems and Biodiversity. Ecological and Economic Foundations. Earthscan, London and Washington.

Ten Brink B (2006) A Long-Term Biodiversity, Ecosystem and Awareness Research Network Indicators as communication tools: an evolution towards composite indicators. ALTER-Net. https://seea.un.org/sites/seea.un.org/files/9_21.pdf

Terraube J, Van Doninck J, Helle P et al. (2020) Assessing the effectiveness of a national protected area network for carnivore conservation. Nature Communications 11:2957. https://doi.org/10.1038/s41467-020-16792-7

Tesfaw AT, Pfaff A, Golden Kroner RE et al. (2018) Land-use and land-cover change shape the sustainability and impacts of protected areas. PNAS 115(9): 2084–2089. https://doi.org/10.1073/pnas.1716462115

Tilman D, May RM, Lehman CL et al. (1994) Habitat destruction and the extinction debt. Nature 371: 65–66. https://doi.org/10.1038/371065a0

Tinaut A, Ruano F (2002) Biodiversidad, clasificación y filogenia. In: Soler M (ed) Evolución: la base de la biología, Manuel Soler, p 293–306.

Trimble M J, van Aarde RJ (2012) Geographical and taxonomic biases in research on biodiversity in human-modified landscapes. Ecosphere 3(12): 119. https://doi.org/10.1890/ES12-00299.1

Troudet J, Grandcolas P, Blin A et al. (2017) Taxonomic bias in biodiversity data and societal preferences. Scientific Reports 7: 9132. https://doi.org/10.1038/s41598-017-09084-6

Trujillo-Miranda AL, Toledo-Aceves T, López-Barrera F, Gerez-Fernández P (2018) Active versus passive restoration: Recovery of cloud forest structure, diversity and soil condition in abandoned pastures. Ecological Engineering 117: 50–61. https://doi.org/10.1016/j.ecoleng.2018.03.011

UN (2021) The UN Decade on Ecosystem Restoration 2021-2030. Strategy. https://www.decadeonrestoration.org/strategy. Accessed 29 June 2021

UNDPA (1972) Report of the United Nations Conference on the Human Environment. Stockholm. https://www.un.org/ga/search/view_doc.asp?symbol=A/CONF.48/14/REV.1. Accessed 11 March 2021

UNDP (2020) Human Development Report 2020. The next frontier. Human development and the Anthropocene. United Nations Development Programme. UNDP, New York. http://hdr.undp.org/sites/default/files/hdr2020.pdf Retrieved 4 Mai 2021

UNEP-FAO (2021) Preventing, halting and reversing the degradation of ecosystems worldwide. The UN Decade on Ecosystem Restoration is a global rallying cry to heal our planet. What will you restore? https://www.decadeonrestoration.org/. Accessed 11 March 2021

UNEP-WCMC (2010) Management Effectiveness (PAME). https://www.protectedplanet.net/en/thematic-areas/protected-areas-management-effectiveness-pame?tab=Results. Accessed 28 June 2021

United Nations (2021) Genocide. Office on Genocide Prevention and the Responsibility to Protect https://www.un.org/en/genocideprevention/genocide.shtml. Accessed 11 March 2021

van Dinter R, Catal C, Tekinerdogan B (2021) A Multi-Channel Convolutional Neural Network approach to automate the citation screening process. Applied Soft Computing 112: 107765. https://doi.org/10.1016/j.asoc.2021.107765

Van Swaay C, Regan E, Ling M. et al. (2015). Guidelines for Standardised Global Butterfly Monitoring. GEO BON, Leipzig, GEO BON Technical Series 1, 32 pp.

Vieilledent G, Cornu C, Cuní Sanchez A et al. (2013) Vulnerability of baobab species to climate change and effectiveness of the protected area network in Madagascar: Towards new conservation priorities. Biological Conservation 166: 11–22. https://doi.org/10.1016/j.biocon.2013.06.007

Vihervaaram P, Auvinen AP, Mononen L et al. (2017) How Essential Biodiversity Variables and remote sensing can help national biodiversity monitoring. Global Ecology and Conservation 10: 43–59. https://doi.org/10.1016/j.gecco.2017.01.007

Wang X, Wang Y, Wang Y (2013) Use of exotic species during ecological restoration can produce effects that resemble vegetation invasions and other unintended consequences. Ecological Engineering 52: 247–251. https://doi.org/10.1016/j.ecoleng.2012.11.007

Wang L, Pan Y, Cao Y et al. (2018) Detecting early signs of environmental degradation in protected areas: An example of Jiuzhaigou Nature Reserve, China. Ecological Indicators 91: 287–298. https://doi.org/10.1016/j.ecolind.2018.03.080

Wanghe K, Guo X, Hu F et al. (2020) Spatial coincidence between mining activities and protected areas of giant panda habitat: The geographic overlaps and implications for conservation. Biological Conservation 247: 108600. https://doi.org/10.1016/j.biocon.2020.108600

Watson J, Dudley N, Segan D et al. (2014) The performance and potential of protected areas. Nature 515:67–73. https://doi.org/10.1038/nature13947

Westengen OT, Jeppson S, Guarino L (2013) Global Ex-Situ Crop Diversity Conservation and the Svalbard Global Seed Vault: Assessing the Current Status. PLoS ONE 8(5): e64146. https://doi.org/10.1371/journal.pone.0064146

White TD, Carlisle AB, Kroodsma DA et al. (2017) Assessing the effectiveness of a large marine protected area for reef shark conservation. Biological Conservation 207: 64–71. https://doi.org/10.1016/j.biocon.2017.01.009

Wilkie DS, Carpenter JF, Zhang Q (2001) The under-financing of protected areas in the Congo Basin: so many parks and so little willingness-to-pay. Biodiversity and Conservation 10: 691–709. https://doi.org/10.1023/A:1016662027017

Wilson EO (2016) Half earth: Our planet's fight for life. Liveright, New York.

Woodruff DS (2001) Populations, species, and conservation genetics. In: Asher, S (ed). Encyclopedia of Biodiversity. Academic Press, p 811–829.

WWF (2020) Living Planet Report 2020-Bending the curve of biodiversity loss. Almond REA, Grooten M, Petersen T (Eds). WWF, Gland. https://www.worldwildlife.org/publications/living-planet-report-2020. Accessed 21 April 2021

WWF (2021) Living Planet Index. WWF, Gland https://wwf.panda.org/discover/knowledge_hub/all_publications/living_planet_index2/. Accessed 25 August 2021

Xu H, Qiang S, Genovesi P et al. (2012) An inventory of invasive alien species in China. NeoBiota 15: 1–26. https://doi.org/10.3897/neobiota.15.3575

Xu W, Pimm SL, Du A et al. (2019) Transforming Protected Area Management in China. Trends in Ecology & Evolution 34 (9): 762–766. https://doi.org/10.1016/j.tree.2019.05.009

Yao RT, Scarpa R, Harrison DR et al. (2019) Does the economic benefit of biodiversity enhancement exceed the cost of conservation in planted forests? Ecosystem Services 38: 100954. https://doi.org/10.1016/j.ecoser.2019.100954

Young HS, McCauley DJ, Galetti M et al. (2016) Patterns, causes, and consequences of Anthropocene defaunation. Annual Review of Ecology, Evolution, and Systematics 47: 333–358. https://doi.org/10.1146/annurev-ecolsys-112414-054142

Zang Z, Zou, X, Zuo P et al. (2017) Impact of landscape patterns on ecological vulnerability and ecosystem service values: An empirical analysis of Yancheng Nature Reserve in China. Ecological Indicators 72: 142–152. https://doi.org/10.1016/j.ecolind.2016.08.019

Index

D. Rodríguez-Rodríguez, J. Martínez-Vega, *Effectiveness of Protected Areas in Conserving Biodiversity*, Strategies for Sustainability,
https://doi.org/10.1007/978-3-030-94297-7

The manufacturer's authorised representative in the EU is Springer
Nature Customer Service Centre GmbH, Europaplatz 3, 69115 Heidelberg,
Germany. If you have any concerns regarding our products, please
contact ProductSafety@springernature.com

Printed and bound by CPI Group (UK) Ltd, Croydon, CR0 4YY
01/10/2026
02225880-0016